CELL BIOLOGY RESEARCH PROGRESS

CELLULOSE BIOSYNTHESIS INHIBITORS AS TOOLS FOR RESEARCH OF CELL WALL STRUCTURAL PLASTICITY

CELL BIOLOGY RESEARCH PROGRESS

Additional books in this series can be found on Nova's website
under the Series tab.

Additional E-books in this series can be found on Nova's website
under the E-book tab.

CELL BIOLOGY RESEARCH PROGRESS

CELLULOSE BIOSYNTHESIS INHIBITORS AS TOOLS FOR RESEARCH OF CELL WALL STRUCTURAL PLASTICITY

JESÚS MIGUEL ÁLVAREZ,
ANTONIO ENCINA,
PENÉLOPE GARCÍA-ANGULO,
ANA ALONSO-SIMÓN,
HUGO MÉLIDA
AND
JOSÉ LUIS ACEBES
EDITORS

Nova Science Publishers, Inc.
New York

Copyright © 2012 by Nova Science Publishers, Inc.

All rights reserved. No part of this book may be reproduced, stored in a retrieval system or transmitted in any form or by any means: electronic, electrostatic, magnetic, tape, mechanical photocopying, recording or otherwise without the written permission of the Publisher.

For permission to use material from this book please contact us:
Telephone 631-231-7269; Fax 631-231-8175
Web Site: http://www.novapublishers.com

NOTICE TO THE READER

The Publisher has taken reasonable care in the preparation of this book, but makes no expressed or implied warranty of any kind and assumes no responsibility for any errors or omissions. No liability is assumed for incidental or consequential damages in connection with or arising out of information contained in this book. The Publisher shall not be liable for any special, consequential, or exemplary damages resulting, in whole or in part, from the readers' use of, or reliance upon, this material. Any parts of this book based on government reports are so indicated and copyright is claimed for those parts to the extent applicable to compilations of such works.

Independent verification should be sought for any data, advice or recommendations contained in this book. In addition, no responsibility is assumed by the publisher for any injury and/or damage to persons or property arising from any methods, products, instructions, ideas or otherwise contained in this publication.

This publication is designed to provide accurate and authoritative information with regard to the subject matter covered herein. It is sold with the clear understanding that the Publisher is not engaged in rendering legal or any other professional services. If legal or any other expert assistance is required, the services of a competent person should be sought. FROM A DECLARATION OF PARTICIPANTS JOINTLY ADOPTED BY A COMMITTEE OF THE AMERICAN BAR ASSOCIATION AND A COMMITTEE OF PUBLISHERS.

Additional color graphics may be available in the e-book version of this book.

Library of Congress Cataloging-in-Publication Data

Cellulose biosynthesis inhibitors as tools for research of cell wall structural plasticity / editors, Jeszs Miguel Alvarez ... [et al.].
       p. cm.
   Includes bibliographical references and index.
   ISBN 978-1-61470-946-6 (softcover)
   1. Cellulose--Inhibitors. 2. Cellulose--Metabolism--Regulation. 3. Plant cell walls--Research--Methodology. I. Alvarez, Jeszs Miguel.
   QK898.C42C445 2011
   572'.56682--dc23
                                                            2011028037

Published by Nova Science Publishers, Inc. † New York

CONTENTS

Preface		**vii**
Chapter 1	**Introduction**	**1**
	Cellulose Biosynthesis	*2*
Chapter 2	**Cellulose Biosynthesis Inhibitors**	**5**
	Dichlobenil	*5*
	Isoxaben	*11*
	Thaxtomin A	*16*
	Other CBIs	*19*
Chapter 3	**Compounds Proposed to Indirectly Inhibit Cellulose Biosynthesis**	**25**
	Quinclorac	*25*
	Ancymidol	*28*
	Coumarin and Derivatives	*29*
	Cobtorin	*32*
	Triaziflam	*33*
	Indaziflam	*34*
	MBTU	*35*
	Oxaziclomefone	*35*
Chapter 4	**CBIs-related Mutants**	**37**
	Mutants Resistant to CBIs	*41*
	Mutants with Changed Sensitivity to CBIs	*43*

	Some Mutants Show Similar Phenotypes to CBIs-Treated Plants	*44*
Chapter 5	**CBIs as Tools to Research the Structural Plasticity of Cell Walls**	**47**
	Habituation to CBIs	*47*
	Habituation to Dichlobenil	*48*
	Habituation to Isoxaben	*50*
	Habituation to Quinclorac	*51*
	Habituation to Thaxtomin A	*51*
	Dehabituation	*51*
Chapter 6	**New Perspectives in CBIs Uses**	**53**
Conclusion		**55**
References		**57**
Index		**79**

PREFACE

This monograph deals with a heterogeneous group of structurally unrelated compounds, called cellulose biosynthesis inhibitors (CBIs). This group includes: dichlobenil, isoxaben, thaxtomin A, flupoxam, triazofenamide, compound 1, CGA 325′615 and AE F150944. Additionally, other putative CBIs that inhibit cellulose biosynthesis in a secondary effect are also considered, such as: quinclorac, ancymidol, coumarin and derivatives, triaziflam, indaziflam, oxaziclomefone and cobtorin. With the exception of thaxtomin A, the only naturally occurring CBI, the rest of CBIs are synthetic compounds. The mode of action of CBIs is mostly unknown; however, some of them are used as herbicides: dichlobenil, isoxaben flupoxam, triaziflam and indaziflam, and are included into group L in the Herbicide Resistance Action Committee classification of herbicides (inhibition of cellulose synthesis).

Cellulose, a linear β-1,4-linked glucan, is a major component of plant cell walls, and is the world's most abundant biopolymer. Despite the importance of cellulose, its synthesis remains largely unknown.

CBIs can be powerful tools for the molecular dissection of cellulose biosynthesis processes; including the assembly between CESA subunits, and the relation between cortical microtubules and cellulose deposition. Moreover, in the last ten years, the isolation of mutants resistant to CBIs has been proved to be a powerful tool to unravel the composition and organization of the cellulose biosynthesis "machinery" of primary cell walls.

Finally, the habituation of cell cultures to CBIs has resulted in useful insight regarding the mechanisms that underlie the plasticity of plant cell wall structure and composition.

Chapter 1

INTRODUCTION

Cellulose biosynthesis inhibitors (CBIs) constitute a group of structurally diverse compounds with different modes of action. The specific site of action of most CBIs is still unknown; however, since they specifically impair certain steps in the cellulose synthesis and deposition pathways in higher plants, they are often used to investigate cellulose synthesis processes. An important set of data regarding CBIs has been raised since two particular reviews were afforded in 1999 (Sabba and Vaughn) and Vaughn (2002); and now the global view of CBIs is more panoramic.

Several CBIs are used as herbicides and are included as group L in the Herbicide Resistance Action Committee (HRAC) classification of herbicides: dichlobenil, isoxaben, chlorthiamid triaziflam and indaziflam. Likewise, several other compounds, such as triazofenamide and triaziflam (Wakabayashi and Böger, 2004), have also been cited as promising herbicides. Until now, however, they have neither been commercialized nor included into this group. Additionally, quinclorac is also considered an auxinic herbicide and is therefore included in the group O (Menne and Köcher, 2007), while oxaziclomefone is still included in those herbicides with unknown mode of action (group Z). Finally, some drugs have been described to display a dual effect (i.e. quinclorac or ancymidol): acting as CBIs in some cases (i.e. depending on the species, or their concentration); and showing an additional mode of action, in other circumstances. The common herbicidal activity of

CBIs is likely caused by separate sites of action that, either directly or indirectly, involves cellulose biosynthesis.

The intention of this report is to describe a global view of CBIs, while paying special regard to their molecular structure and mode of actions. Furthermore, this report will address CBIs-related mutants and their usefulness as research tools in investigating both cellulose synthesis and deposition in the primary cell wall. Lastly, habituation to CBIs in cell cultures and studies which reflect the plasticity of plant cell walls are also described.

CELLULOSE BIOSYNTHESIS

Cellulose is the most abundant organic compound on earth, has a high physiological importance, and many applications in the industry. However, many aspects of its biosynthesis still remain unknown; and until recently, have we started to unravel some details of this process. Cellulose is formed from unbranched β-1,4-linked chains of glucose residues, aggregated to form microfibrils. Due to the high mechanical strength of these microfibrils, their oriented deposition in the cell wall is essential for directional cell growth.

Since several reviews on cellulose synthesis have been reported in recent years (Somerville, 2006; Joshi and Mansfield, 2007; Mutwil et al., 2008; Taylor, 2008; Bessueille and Bulone, 2008; Guerriero et al., 2010; Harris et al., 2010; Carpita, 2011), only a brief presentation of cellulose biosynthesis will be introduced at this time. Globally, cellulose microfibril formation can be divided into three steps: i) initiation, using UDP-glucose as the donor substrate; ii) polymerization of glucose into β-1,4-glucan chains; and iii) crystallization of β-1,4-glucan chains into a microfibril (Peng et al., 2002). Based on *in vitro* experiments, it has been proposed that cellulose biosynthesis is initiated from a sitosterol-β-glucoside as primer molecule (Peng et al., 2002).

In higher plants, cellulose is synthesized by transmembrane complexes called terminal complexes, or rosettes, which can be observed as hexamers; using freeze-fracture electron microscopy. Notably, the purified rosettes did not possess cellulose synthesis activity *in vitro*. Rosette complexes consist of 36 individual cellulose synthase (CESA) proteins (six CESA proteins per each monomer) that are thought to be the catalytic subunits of the cellulose synthase complexes (Doblin et al., 2002; Somerville, 2006). These proteins are encoded by a family of CesA genes that are found throughout the plant kingdom (Richmond and Somerville, 2000; Hazen et al., 2003).

At least three different CESA proteins, encoded by members of the CESA gene family, are required for cellulose biosynthesis in primary cell walls (Taylor, 2008): CESA1 and CESA3, combined with a member of the CESA6-like clade (CESA2, CESA5, CESA6, or CESA9) (Desprez et al., 2007; Persson et al., 2007). In *Arabidopsis thaliana* these proteins are integrated in CESA complexes in such a way that CESA3 and CESA6 physically interact (Desprez et al., 2007).

It is assumed that the three different CESA proteins cluster into the rosette structure, forming the six subunits of the hexamer within the complex. Each subunit makes six glucan chains, resulting in the synthesis of 36 glucan chains per rosette complex. Additional proteins seem to be necessary for several functions: as the formation of primers, metabolic channeling of substrates, and crystallization and further termination of chains (Guerriero et al., 2010; Carpita, 2011). Studies with fluorescent proteins have elucidated that the assembly of CESA complexes takes place in the Golgi apparatus and then are exported, via exocytosis, to the plasma membrane (Paredez et al., 2006; Somerville, 2006; Desprez et al., 2007).

Genetic experiments provide evidence that additional proteins form part of the complex: a cytoskeletal-anchored sucrose synthase, that could channel UDP-glucose to CESA (Amor et al., 1995); a membrane-bound endoglucanase/cellulase, that is thought to act as an editing/repairing protein during cellulose biosynthesis (Molhoj et al., 2002); a plasma membrane protein, denominated KOBITO (Pagant et al., 2002); and COBRA, a GPI-anchored protein (Schindelman et al., 2001). Recently, the first non-CESA protein which exerts a direct effect on cellulose synthesis, through its association with CESA complexes, has been described (Gu et al., 2010). Mutations in *Csi1* affect the distribution and movement of CESA complexes, resulting in strongly reduced rates of CESA complex movement. CSI1 may participate in the mechanisms that are responsible for organizing the deployment of cellulose microfibrils in primary walls. However, the precise role of this and roles of the other proteins are still not clear.

Another set of three CESA proteins is required for cellulose synthesis in secondary cell walls: CESA4, CESA7 and CESA8 (Atanassov et al., 2009; Timmers et al., 2009). In this case, purified CESA complexes did not contain any further protein (Atanassov et al., 2009). However, it has been recently proposed that two genes *Tbr* (AT5G06700) and *Tbr-like3* (*Tbl3*) (AT5G01360) are responsible for the trichome birefringence trait; and encode proteins that contribute to the synthesis and deposition of secondary cellulose, presumably by influencing the esterification state of the pectic polymer (Bischoff et al., 2010).

Cortical microtubules determine the orientation of the cellulose microfibrils that are deposited in the cell wall, through the regulation of the dynamics of cellulose synthase complexes, by three mechanisms: targeting their secretion through interactions with the Golgi apparatus (Crowell et al., 2009; 2010), guiding their movement through the plasma membrane (Paredez et al., 2006), and mediating the internalization of cellulose synthase complexes into cortical microtubule-associated compartments (Gutierrez et al., 2009).

Chapter 2

CELLULOSE BIOSYNTHESIS INHIBITORS

DICHLOBENIL

Dichlobenil (2,6-dichlorobenzonitrile) is the simplest and more studied CBI. As a consequence of microorganism metabolism (Beynon and Wright, 1968), a related herbicide, chlorthiamid (2,6-dichlorothiobenzamide), is converted to dichlobenil in soil. First registered as a pesticide in the U.S. in 1964, dichlobenil was originally discovered in the 1950's by Phillips-Dumar, in the Netherlands. Traditionally, it has been used as a broad spectrum preemergence herbicide: to control biennial and perennial weeds in agricultural, residential and industrial areas; to remove tree roots and inhibit their growth in sewers; and, finally, in the capacity as an aquatic herbicide. Today, it continues to be commercially available under such common names as: Prefix, Barrier, Casoron, Dyclomec, Norosac.

Dichlobenil is degraded mainly to 2,6-dichlorobenzamide by biotic or non-biological processes in the hydrosoil (Beynon and Wright, 1968; Montgomery et al., 1972; Verloop, 1972). Having been positively measured five years after application, both 2,6-dichlorobenzamide and dichlobenil, are "remarkably persistent" in the soil. Additionally, dichlobenil volatilizes (vaporizes) readily, resulting in possible contamination of the air in areas where it has been used.

Dichlobenil

Dichlobenil is acutely toxic to animals: it affects reproduction in both male and female animals, causes cancer (the U.S. Environmental Protection Agency classifies dichlobenil as a possible human carcinogen), and is highly toxic for the olfactory mucosa (Brittebo et al., 1991). The oral LD_{50} for acute toxicity of dichlobenil to mammalians varied between 2056 and 3160 mg.kg^{-1} body weight (Brooker and Edwards, 1974 and refs. therein); and the LC_{50} range for acute toxicity of dichlobenil to fish varied between 14.5 mg.l^{-1} for 24 h of exposition and 12.3 mg.l^{-1} for 48 h. Dichlobenil reduces the growth of beneficial mycorrhizal fungi on the roots of apple trees (Pedersen and Sylvia, 1996). Due to these adverse reactions, dichlobenil is banned in Denmark, amongst other countries.

Dichlobenil is absorbed by the roots and is slowly transported upwardly in xylem with minimal to no downward movement in phloem (Verloop and Nimmo 1969, 1970). It acts primarily on growing points and root tips, and is a powerful inhibitor of both seed germination and actively dividing meristems. Dichlobenil has a stronger consequence on seedlings, through impairing their growth; than it has on the actual inhibition of seed germination (eg. radicle growth of French bean is 30 times more sensible to dichlobenil than seed germination (Encina, unpubl.).

I_{50} values in the micromolar range have been measured for the affect of dichlobenil on root growth: 1 µM on *Lepidium sativum* (Günther and Pestemer, 1990), 0.4 µM on Arabidopsis (Heim et al., 1998), 4 µM on the French bean (Encina, unpubl.), and 2 µM on maize (Mélida, unpubl.). Rapidly expanding cells; such as suspension or callus-cultured cells (Shedletzky et al., 1990, 1992; Corio-Costet et al., 1991a, b; Encina et al., 2001, 2002; Mélida et al., 2009), seedling roots and hypocotyls (Himmelspach et al., 2003; DeBolt et al., 2007b), and pollen tubes (Anderson et al., 2002), are sensible to dichlobenil in the nano-micromolar range (I_{50}: 50 nM for soybean suspension-cultured cells -Corio-Costet et al., 1991b-; 0.5 µM and 0.3 µM for French bean callus and suspension-cultured cells respectively -Encina et al., 2001, 2002- and 1.5 µM for maize callus -Mélida et al., 2009-).

Dichlobenil inhibits the regeneration of cell walls in protoplasts (Meyer and Herth 1987), formation of cell plates (Buron and Garcia-Herdugo 1983), extension of root hairs and secondary-root development (Meekes 1986), and the formation of cellulose microfibrils (Mizuta and Brown 1992). Moreover, it provokes radial root or hypocotyl swelling (Umetsu et al., 1976; Eisinger et al., 1983; Montague, 1995; Himmelspach et al., 2003; DeBolt et al., 2007b) and induces necrotic lesions (Barreiro, unpubl.). These symptoms are consistent with an impairment of some stages of cellulose biosynthesis. Thus, since it was observed that dichlobenil impairs the incorporation of radiolabelled glucose into cellulose (Hogetsu et al., 1974); the inhibition of cellulose biosynthesis has, as a result, been later demonstrated in a wide range of systems (Montezinos and Delmer, 1980; Brummell and Hall, 1985; Hoson and Masuda, 1991; Corio-Costet et al., 1991b; Edelmann and Fry, 1992; Shedletzky et al., 1992; García-Angulo et al., 2009a). Since the synthesis of other cell wall polysaccharides is not affected after a short term dichlobenil treatment (Montezinos and Delmer, 1980; Blaschek et al., 1985; Francey et al., 1989), the effect of dichlobenil on cellulose biosynthesis appears to be specific. Furthermore, dichlobenil has additionally been reported to inhibit the non-cellulosic polysaccharides synthesis in algae (Arad et al., 1994; Wang et al., 1997).

Apart of the inhibition of glucose incorporation into cellulose, it has also been suggested to be a side effect of dichlobenil on cellulose microfibril orientation (Sugimoto et al., 2001), cellulose-synthesizing complexes organization/motility (Herth, 1987; Mizuta and Brown, 1992; DeBolt et al., 2007b; Wightman et al., 2009), microtubule organization (Himmelspach et al., 2003) and callose synthesis (Vaughn et al., 1996; DeBolt et al., 2007a; Apostolakos et al., 2009).

Habituation of plants for growth in the presence of dichlobenil reveals an accumulation of noncrystalline β-1,4-glucan (Encina et al., 2002; Garcia-

Angulo et al., 2006). The effect of dichlobenil is fully reversible, and the removal of dichlobenil from the growth medium induces patches in the plasma membrane that accumulate fibrilar material (Garcia-Angulo et al., 2006). Interestingly, the accumulation of this material seems to coincide with regions of close contacts with the plasma membrane and cortical microtubules (Meyer and Herth, 1978). In algae, dichlobenil seems to interfere with the assembly of the linear terminal complexes implicated in cellulose biosynthesis (Mizuta and Brown, 1992). Furthermore, dichlobenil treatments have been shown to cause changes in number of intact rosettes at the plasma membranes of algae and wheat (Herth, 1987; Rudolph et al., 1989) and accumulation of cellulose synthase subunits within localized regions at the plasma membrane of Arabidopsis hypocotyls (DeBolt et al., 2007b). Recently, dichlobenil treatments have revealed to cause cessation of the CESA mobility at the plasma membrane of Arabidopsis (DeBolt et al., 2007b; Crowell et al., 2010). A common theme between these different observations seems to be that dichlobenil inhibits cellulose synthesis by disrupting the ordered deposition of crystalline cellulose; rather than inhibiting the polymerization of the β-1,4-glucan chains, *per se*.

Mode of Action

Aside that it was discovered more than fifty years ago and used as a specific CBI for decades, the mode of action of dichlobenil is largely unknown.

Initially, short-term effects of dichlobenil on cellulose synthesis were observed in freeze-fracture electron micrographs. It was shown that the addition of dichlobenil to wheat (*Triticum aestivum*) roots had stimulated the accumulation of more than double the number of CESA rosettes in the plasma membrane (Herth, 1987). Later, by using live-cell imaging of transgenic plants carrying a yellow fluorescent protein (YFP)-labeled cellulose synthase 6 (CESA6) protein (DeBolt et al, 2007b), it was suggested that dichlobenil rapidly inhibited the mobility of CESA6 protein at the cortical z-plane in Arabidopsis cells and caused hyperaccumulation of CESA complexes at discrete sites at the cell cortex. As it was suggested by DeBolt et al. (2007b); the microtubule motors, such as kinesins, are unlikely to be the engine for

CESA motility; rather, CESA movement is likely propelled by the polymerization force of cellulose synthesis.

Based on the hypothesis that sitosterol-β-glucoside is the primer for CesA glucosyltransferase to initiate glucan polymerization (Peng et al., 2002), a mode of action proposed for dichlobenil consists on blocking the initiation steps of cellulose biosynthesis. Evidences for such idea are: i) dichobenil inhibited the *in vivo* synthesis of sitosterol-β-glucosides; ii) dichlobenil reduced the incorporation of radioactive glucose into sterol in isolated cotton (*Gossypium hirsutum*) fibers (Peng et al., 2002), either by the inhibition of the formation of UDP-glucose or through some other indirect effect; and iii) the exogenous addition of sitosterol-β-glucoside has the capacity to reverse the inhibition properties of dichlobenil. However, due to the seemingly highly forced experimental conditions used by Peng et al. (2002), this mode of action for dichlobenil is now considered controversial. As a result, hypothesis that cellulose synthesis is initiated from a primer is not well accepted and should be further tested (Somerville, 2006). Interestingly, it was described that *in vitro* cellulose synthesis did not require addition of any primer. Furthermore, no difference in CESA6 mobility was detected at all between dichlobenil alone and dichlobenil plus sitosterol-β-glucoside (DeBolt et al., 2007a).

A second mode of action proposed for dichlobenil stems from the observation that a severe reduction in the synthesis of cellulose alters orientation of the remaining cellulose microfibrils. This alteration was observed in the cellulose-deficient elongating cells of *kob* (Pagant et al., 2002) and *rsw1* (Sugimoto et al., 2001) Arabidopsis mutants; and in the wild type of Arabidopsis cells treated with 1 µM dichlobenil, where cellulose microfibrils were randomly oriented. Therefore, it has been suggested, that the dichlobenil effect on cellulose biosynthesis would result in an alteration of cellulose crystallization; rather than of an inhibition of glucose polymerization. Since it was observed that dichlobenil disrupted the orientation of microtubules in Arabidopsis root epidermal cells (Himmelspach et al., 2003), it is probable that this effect occurs through microtubule organization. In this respect, additionally remarkable, is that the reduced level of cellulose in cell walls from dichlobenil-habituated cells (as it will be detailed further) is accompanied by an accumulation of a non-crystalline β-1,4-glucan (Encina et al., 2002; García-Angulo et al., 2006).

Presently, three putative targets for dichlobenil have been reported. A putative dichlobenil receptor of 18 kDa in plants was characterized using a photoreactive analog of dichlobenil (2,6-dichlorophenylazide) (Delmer et al., 1987). The amount of this polypeptide seemed to increase significantly at the onset of secondary cell wall cellulose synthesis in the fibers. The polypeptide that is loosely associated with membranes is unlikely to be the catalytic subunit of the cellulose synthase; although it may function as a regulatory protein for β-glucan synthesis.

The second molecule reported to act as a receptor for dichlobenil was the CESA1 protein. This idea was based on the the specific binding of dichlobenil to the CESA1 protein, observed by Nakagawa and Sakurai (1998). Moreover, the *rsw1* mutant was isolated as a mutant that mimics responses of wild type roots to dichlobenil, resulting in the reduction of the cellulose content and accumulation of a non-crystalline β-1,4-glucan. As the *Rsw1* locus of Arabidopsis encodes the catalytic subunit of cellulose synthase (CESA1), the cloned gene complements the *rsw1* mutant (Arioli et al., 1998). Therefore, it is likely a common mechanism between the action of dichlobenil and *rsw1* mutation.

Long ago, it was demonstrated that dichlobenil stimulates the accumulation of CESA complexes in the plasma membrane of wheat (Herth, 1987). More recently, live-cell imaging of transgenic plants, carrying a yellow fluorescent protein (YFP)-CESA6 fusion, showed that a short-term treatment with dichlobenil inhibited the motility of these complexes in Arabidopsis cells and also promoted their hyperaccumulation at sites in the plasma membrane that may coincide with loading areas of CESA complexes from Golgi (DeBolt et al., 2007b). Further studies confirmed that dichlobenil additionally slowed the movement of CESA complexes beneath the zones of formation of secondary wall (Wightman et al., 2009). These findings may reveal the interference of dichlobenil with the circulation of CESA complexes between Golgi and the plasma membrane.

A microtubule, associated protein MAP20 in secondary cell walls of hybrid aspen, was recently reported as a target for dichlobenil (Rajangam et al., 2008). This finding was preceeded by the observation that dichlobenil induced changes in the cortical microtubule networks in the fucoid algae, *Pelvetia compressa* (Bisgrove and Kropf, 2001), and Arabidopsis roots (Himmelspach et al., 2003). MAP20 is a small cytosolic protein strongly upregulated during the formation of the secondary cell wall and shares a conserved domain with a classical microtubule associated protein, TPX2, demonstrated to bind microtubules and proteins that use microtubules as guiding templates. MAP20 binds to microtubules, both *in vitro* and *in vivo*. During cellulose synthesis, dichlobenil specifically binds to MAP20, but it does not prevent the binding of the protein to microtubules. The authors propose that, in developing secondary walls, MAP20 has a role in coupling the "machinery" of cellulose synthesis with cortical microtubules; and that dichlobenil inhibits cellulose biosynthesis by decoupling cellulose synthesis and microtubules through MAP20 inactivation. Finally, it is suggested that the small polypeptide described by Delmer et al. (1987) might be a putative ortholog of MAP20 (Rajangam et al., 2008).

Recently, studies on the expression profile of *ZmCesA* genes in dichlobenil-habituated maize cells have indicated that the abscense of ZmCESA5 within the rossete, could be a key feature in determining the tolerance of habituated cells in coping with dichlobenil (Mélida et al., 2010b). Although these results do not constitute a direct proof for a dichlobenil target in maize cells; the possibility for the ZmCESA5 protein to act (primarily or secondary) as dichlobenil target, should be considered.

ISOXABEN

Isoxaben (N-[3-(ethyl-1-methylpropyl)])-5-isoxazolyl-2,6 dimethoxy-benzamide is a pre-emergence herbicide, available commercially as Flexidor, Gallery or EL-107. Isoxaben is a white crystalline solid; soluble in organic solvents like acetone, acetonitrile and dichloromethane. Generally classified as nontoxic on mammalians, it has an oral LD50 in rats of more than 10,000 mg kg-1. Additionally, isoxaben has a moderate chemical persistence and relatively low mobility in soil. While it is metabolized within the soil into nontoxic products, organic fertilizer treatments simultaneously increase the soil persistence of isoxaben (Rouchaud et al., 1993). Due to its selectively phytotoxic to dicotyledonous plants (most monocotyledonous species are tolerant); isoxaben is marketed for: season-long control of annual broadleaf weeds in winter-cereal crops, established turf, ornamentals and landscape plantings.

Isoxaben is very active in the nanomolar range: the seedling growth is 50% inhibited at 1.5 nM (Desprez et al., 2002) and 4.5 nM (Heim et al., 1989) in Arabidopsis, and at 20 nM (Lefebvre et al., 1987) in *Brassica napus*. Furthermore, isoxaben very actively inhibits the growth of *in vitro* cultured-cells. I_{50} values in the nanomolar range have been reported for different systems: 170 nM for Arabidopsis calluses (Heim et al., 1989), 80 nM for soybean cell suspensions (Corio-Costet et al., 1991a, b), and 10 nM for French bean calluses (Díaz-Cacho et al., 1999).

Injury that is caused by isoxaben on weeds includes both root and shoot effects (Schneegurt et al., 1994; Sabba and Vaughn, 1999). The primary response of susceptible plants to isoxaben, when it is applied to soil, is a swelling of the seedling's root tip. Additionally, other root transormations might include: nubbing, stunting and/or discoloration. Likewise, changes in the aerial part of the plant include both swollen and split stems and petioles, and formative effects on leaves. Leaf browning, curled leaves and meristematic shoot death were also seen, namely in plants with little cuticle development.

Isoxaben

Corio-Costet *et al.* (1991b) reported that cytological symptoms seen in tissue-cultured cells of a sensitive soybean [*Glycine max* (L.) Merr.] included a detachment of the plasma membrane from the cell wall and the deposition of fibrilar material in the extracytoplasmic space. These symptoms, indicative of some abnormality of the cell wall, were absent in tolerant cell cultures of the soybean (Corio-Costet *et al.*, 1991b).

Isoxaben symptomatology evokes that of a dichlobenil. Arabidopsis seedlings, treated with isoxaben, show a typical dwarf phenotype; caused by the inhibition of hypocotyl and root elongation. Treated cells of sensitive species fail to elongate normally and, consequently, grow isodiametrically (Lefebvre et al., 1987). Therefore, isoxaben-treated organs typically expand radially and have reduced elongation rates. Moreover, isoxaben causes an accumulation of callose and ectopic lignification (Desprez et al., 2002; Caño-Delgado et al., 2003). As a dichlobenil, isoxaben also arrests cell plate formation (Samuels et al., 1995; Vaughn et al., 1996; Durso and Vaughn, 1997). In this respect, isoxaben seems to act in a different manner to dichlobenil, as it has been reported that cell plates of isoxaben-treated BY-2 tobacco cells are reduced in both cellulose and callose; whereas dichlobenil treatment only affected cellulose biosynthesis. In accordance with this and due to the inhibition of cell plate formations at an early stage (Durso and Vaughn, 1997; Sabba and Vaughn, 1999), the effect of isoxaben is more pronounced. Later, by inducing the expression of *Pmr4*, a pathogen- or wound-induced, callose synthase gene (Nishimura et al., 2003); DeBolt et al. (2007b) demonstrated that both dichlobenil and isoxaben promoted the formation of callose on Arabidopsis seedlings

Without interfering with other metabolic processes such as photosynthesis or respiration, or with the synthesis of fatty acids, nucleic acids or pigments (Lefebvre et al., 1987; Heim et al., 1989, 1990; Corio-Costet et al., 1991b; Caño-Delgado et al., 2003); isoxaben specifically inhibits the glucose incorporation into cellulose in plants. The reported I_{50} values for inhibiting the incorporating of [^{14}C]glucose into cellulose are also in the nanomolar range: 10 nM for Arabidopsis (Heim et al., 1990) and 40 nM for soybean (Corio-Costet et al., 1991b). Although preliminary results would indicate an effect of isoxaben on protein synthesis (Lefebvre et al., 1987); more accurate experiments demonstrate that the inhibition of protein synthesis, if produced, was not a direct effect of isoxaben (Heim et al., 1990).

Mode of Action

The target site of isoxaben, identified by using a genetic approach, has shown to be a cellulose synthase (Scheible et al., 2001). Early studies reported that two mutations of Arabidopsis, termed *ixr1* and *ixr2*; with alterations at two genetic loci, were 300 and 90 times, respectively, more resistant to isoxaben than wild-type plants (Heim et al., 1989, 1990). Since resistant cell lines show no alterations in uptake or detoxification of the herbicide (Heim et al., 1991), the *ixr* mutations appear to directly affect the herbicide target. Isolation of the *Ixr1* and *Ixr2* gene, by map-based cloning, revealed that they encode the CESA3 and CESA6 isoforms of cellulose synthase (Scheible et al., 2001; Desprez et al., 2002). No isoxaben-resistant mutants affecting CESA1 (the third "Musketeer" required for cellulose biosynthesis during primary cell wall formation) have been identified, suggesting that CESA1 is not a target for isoxaben (Robert et al., 2004).

The *ixr* mutations occur near the carboxyl terminus, a highly conserved region of the enzyme which is also well separated from the proposed active site, making difficult for isoxaben to interfere directly with the catalytic site of the protein (Desprez et al., 2002). Most probably, *ixr* mutants would become isoxaben-insensitive by a conformational change of the target protein.

The way by which isoxaben prevents the incorporation of glucose into the cell wall is not fully known. It has been shown that isoxaben inhibits cellulose synthesis by perturbing cellulose synthase complexes through the recognition, directly or indirectly, of an epitope which is associated with the complexes containing CESA3 and CESA6, probably located in the interaction of both proteins (Scheible et al. 2001; Desprez et al. 2002, 2007). In particular, the application of isoxaben leads to depletion of CESA6-containing complexes from the plasma membrane (Paredez et al., 2006; DeBolt et al., 2007b). The most probable cause for this isoxaben-specific effect would be the sequestration of CESA in cytoplasmic reservoirs, mediated by a disruption in the motility of compartments involved in the trafficking of these proteins (Gutierrez et al., 2009).

An alternative model explains that an isoxaben-induced conformational change on CESA proteins would block the putative membrane channel needed for the extrusion of the growing cellulose microfibril (Desprez et al., 2002).

As for other CBIs and mutations reported above, the specific inhibition of cellulose biosynthesis by isoxaben results in decreased anisotropy of cell expansion. This effect is carried out through the disorganization of the cortical microtubules in elongating conifer pollen tubes (Lazzaro et al., 2003) and in newly formed *Nicotiana tabacum* protoplasts (Fisher and Cyr, 1998). Paredez et al. (2006) observed that cellulose synthase complexes containing CESA6 are organized in the cell membrane by a functional association with cortical microtubules. The influence of proteins involved in cell wall biosynthesis on cortical microtubule stability and orientation is correlated with cellulose synthesis, rather than the integrity of the cell wall (Paredez et al., 2008). Therefore, the effect of isoxaben on cytoskeleton disorganization could be mediated by the binding of the inhibitor to a site in the CESA6 protein.

Perturbations in the structure or composition of the cell wall, the first barrier against pathogen attacks and environmental challenges, must activate the appropriate response for plant defense. Consequently, it has been shown that an isoxaben treatment triggered ectopic lignin deposition, and defense responses through the ethylene and jasmonate pathways (Ellis et al., 2002; Caño-Delgado et al., 2003; Bischoff et al., 2009). An application of isoxaben (like thaxtomin A, another cellulose biosynthesis inhibitor) on Arabidopsis suspension cells triggered programmed cell death (Duval et al., 2005), which is often associated with the activation of defense mechanisms in plants (Greenberg and Yao, 2004). Both inhibitors activate the expression of a common set of stress-related genes that lead to the programmed cell death (Duval et al., 2009).

Recently observed in Arabidopsis seedlings, isoxaben has activated the *Uxe4* gene (Hamann et al., 2009), which is thought to encode a UDP-D-xylose 4-epimerase, required for the production of arabinose (Seifert, 2004). The activation of *Uxe4* is correlated both with increases in arabinose and uronic acids and also with reductions in galactose and cellulose. The authors proposed that hexoses function as signals in cell-wall stress responses (Hamann et al., 2009). Furthermore, isoxaben upregulated the expression of a putative encoding glycosyltransferase that is similar to the xylosyltransferases involved in the biosynthesis of rhamnogalacturonan II (Fangel et al., 2011)

THAXTOMIN A

Thaxtomin A (a 4-nitroindol-3-yl containing 2,5-dioxopiperazine) is the main phytotoxin produced by *Streptomyces scabiei*, the causative agent of common scab disease (King et al., 1992; King and Calhoun, 2009). At least five other less common Streptomyces species: *Streptomyces acidiscabies* (Lambert and Loria, 1989), *Streptomyces turgidiscabies* (Miyajima et al., 1998), *Streptomyces europaeiscabiei* and *Streptomyces stelliscabiei* (Boucheck-Mechiche et al., 2000) and *Streptomyces niveiscabiei* (Park et al., 2003) are likewise capable of producing thaxtomins.

Recently the protein NEC1, a novel secreted protein required for colonization of the plant host that is thought to suppress plant defense responses during infection was also identified as virulence determinant of *Streptomyces scabies* for suppression of plant defense responses (Bukhalid and Loria, 1997). Moreover, comparative genomic analyses support the involvement of multiple genetic factors in Streptomyces disease development (Bignell et al., 2010).

Thaxtomin A causes substantial wilting in several species after postemergence applications, a symptom dissimilar to that caused by known CBIs (King et al., 2001). Most of the symptoms, however, caused by thaxtomin A are similar to those of dichlobenil and isoxaben (Delmer and Amor, 1995; King et al., 2001; Desprez et al., 2002; Fry and Loria, 2002; Scheible et al., 2003).

At nanomolar concentrations, thaxtomin causes a dramatic swelling of Arabidopsis cells and also a tickening of roots or shoots, due to cell hypertrophy (Scheible et al., 2003). At micromolar concentrations, thaxtomin A (1–3 µM) inhibited normal cell elongation of tobacco protoplasts in a manner that suggested an effect on primary cell wall development (Fry and Loria, 2002). At higher concentrations, similar to those found in scab lesions on field infected potato tubers (Lawrence et al., 1990; Leiner et al., 1996), thaxtomin A provoked cell death. The I_{50} value for thaxtomin A on seedling growth of Arabidopsis ranged between 25 and 50 nM (Scheible et al., 2003).

Recently, by means of a cell culture approach, thaxtomin A has been used to select scab disease-resistant potato plants. This inhibitor was used as a selection agent applied to potato cells culture media: the surviving variants were recovered and used to regenerate complete plants (Wilson et al., 2009).

Thaxtomin A

Mode of Action

The mode of action of thaxtomin A is not known, although this compound has been shown to inhibit cellulose synthesis (Fry and Loria, 2002; Scheible et al., 2003) in a similar manner to that of other CBIs, such as dichlobenil and isoxaben (King et al., 2001; Scheible et al., 2003). In fact, the reduction of seedling growth was accompanied by a reduction of the incorporation of [^{14}C]glucose into the cellulosic fraction of dark-grown Arabidopsis seedlings. This paralleled a significant increase in the incorporation into pectins and slighter increase in the incorporation into hemicelluloses (Scheible et al., 2003; Bischoff et al., 2009). Additional evidence for the inhibition of cellulose biosynthesis was obtained with Fourier transform infrared (FTIR) microspectroscopy. FTIR spectra of thaxtomin A-treated hypocotyls had clustered tightly with those of wild-type hypocotyls treated with CBIs (e.g., isoxaben or dichlobenil) and also with mutants known to be defective specifically in cellulose synthesis (e.g., *rsw1-2* and *kor-2*) (Robert et al., 2004).

The modifications in cell wall composition caused by thaxtomin A were accompanied by changes, both in the expression of *CesA* genes and also in additional cell wall related genes; both in primary (*Korrigan* and *Kobito1*) and secondary (*CesA7, CesA8, Cobra-like 4, Irx8, and Irx9, Cad9*) cell wall synthesis (Bischoff et al., 2009). The alteration in the expression of *CesA* genes of Arabidopsis seedlings result in a depletion of CESA complexes from the plasma membrane, which coincides to their accumulation in a microtubule-associated compartment (Bischoff et al., 2009).

After a treatment with thaxtomin A, changes in the expression of genes which are associated with pectin metabolism and cell wall remodeling were detected. Such was the case of a pectin acetylesterase and two pectin methylesterases that were found to be upregulated by thaxtomin A (Bischoff et al., 2009). The latter result agrees with previous data on compensation of the loss of cellulose by an increased amount of pectin, combined with a lower degree of esterification (Burton et al., 2000).

Thaxtomin A stimulates H^+ efflux across the plasma membrane and a short-lived Ca^{2+} influx in the roots of different species (Tegg et al., 2005). The exchange of Ca^{2+} with H^+ in the apoplast leads to an acidification of the cell wall which would be responsible for enzyme activation and disruption of cellulose synthesis. Therefore, the Ca^{2+} influx was positionated as an upstream event of the inhibition of cellulose synthesis.

Thaxtomin A induces in a wide variety of plant species and tissues; and in a concentration-dependent manner, a cell death that displays programmed cell death features (Duval et al., 2005; Meimoun et al., 2009). Programmed cell death involves fragmentation of nuclear DNA and requires active gene expression and de novo protein synthesis. The Ca^{2+} influx, induced by thaxtomin A, was also reported to be necessary to achieve programmed cell death. This is also true of other early thaxtomin A-induced responses (anion current increase, alkalization of the external medium, and the expression of PAL1), that would appear as a consequence of the inhibition of cellulose biosynthesis (Errakhi et al., 2008). It was recently demonstrated that this programmed cell death occurs by the activation of common stress-related pathways that would somehow bypass the classical hormone-dependent defense pathways (Duval and Beaudoin, 2009).

Sublethal concentrations of several auxins reduced the severity of scab disease (McIntosh et al., 1985) by enhancing tolerance to thaxtomin A (Tegg et al., 2008). Although the mechanism of auxin inhibition of thaxtomin A toxicity is not understood, it could possibly be related to the reversion of programmed cell death by auxins (Gopalan, 2008); or to direct competition for a putative cellular binding, since several auxins share chemical similarities to the thaxtomin A molecule (Tegg et al., 2008).

Finally, the modification of cell wall composition caused by a thaxtomin A treatment, results in an additional cell wall reinforcement by triggering ectopic lignification by a high up-regulation of several genes involved in lignin biosynthesis (Bischoff et al., 2009). Thaxtomin A treatment also provoked the induction of a set of defense genes (Caño-Delgado et al., 2003, Bischoff et al., 2009).

OTHER CBIS

Triazol Carboximide Herbicides (Flupoxam and Triazofenamide)

Flupoxam (1-[4-chloro-3-(2,2,3,3,3-pentafluoropropoxymethyl) phenyl]-5-phenyl-1H-1,2,4-triazole-3-carboximide) and triazofenamide (1- [3-methyl phenyl]-5-phenyl-1H-1,2,4-triazole-3-carboximide) are triazole-carboximide herbicides. Flupoxam, commercialized as Quatatim or KNW-739, is used for broad-leaved weed-control in winter cereals (O'Keefe and Klevorn, 1991), as a preemergence and postemergence herbicide. Flupoxam inhibits the root growth of watercress (*Lepidium sativum*) by 50% at a concentration of 6 nM. Since flupoxam induces classic club root morphology, it was initially characterized as a mitotic disrupter (O'Keefe and Klevorn, 1991). Hoffman and Vaughn (1996), however, later reported that the effect of flupoxam on watercress roots was different than that originally expected of a mitotic disrupter; although they did not propose an alternative mode of action. The treatment of cotton fibers with flupoxam (and also with isoxaben, dichlobenil and thaxtomin A) causes spherical shapes and frequently induces cell division. Fibers which were grown in the presence of isoxaben or flupoxam replaced the entire cell wall with a pectin sheath, chiefly of deesterified pectins; indicating that both herbicides effectively disrupt cellulose biosynthesis and cause radical changes in cell wall structure and composition (Vaughn and Turley, 2001).

Etiolated Arabidopsis seedlings, treated with flupoxam, display a FTIR spectral phenotype that is most closely related to those of either Arabidopsis *rsw1/cesa1* mutant seedlings or Arabidopsis seedlings treated with cellulose synthesis inhibitors like thaxtomin A, dichlobenil or isoxaben (Scheible et al., 2003; Robert et al., 2004). Based on these results, the target molecule of these inhibitors is thought to be CESA1, although further analysis must be done in order to be fully accepted.

Flupoxam   Triazofenamide

The close analog of flupoxam, triazofenamide, was also initially considered as a microtubule polymerization inhibitor (O'Keefe and Klevorn, 1991). Using staining and microscopic techniques, however, this mode of action was later rejected (Hoffman and Vaughn, 1996). Due to the symptoms elicited in an Arabidopsis short-term test, Heim et al. (1998) had postulated that triazofenamide was a cellulose biosynthesis inhibitor. Furthermore, triazofenamide inhibits [^{14}C]glucose incorporation into cellulose in a manner similar to isoxaben (Heim et al., 1998). Nevertheless, the exact modes of action of the triazole-carboximide herbicides are still unknown.

Compound 1

Thiazolidinones, such as Compound 1 (5-tert-butyl-carbamoyloxy-3-(3-trifluromethyl) phenyl-4- thiazolidnone), are a class of N-phenyl-lactam-carbamate herbicides (Sharples et al., 1998). Thiazolidinones show potential for selective preemergence control of a range of weed species in soybean and other crops. Susceptible weeds include grasses: *Digitaria spp.*, *Setaria spp.*, *Sorghum spp.*, *Brachiaria spp.* and *Echinochloa crus-galli* and small-seeded broad-leaved weeds which include *Amaranthus spp.* and *Chenopodium spp.*

Large-seeded, broad-leaved weeds such as *Ipomoea spp.* and *Abutilon spp.* are not controlled, however.

Compound 1

Compound 1 induces a potent (I_{50}= 50 nM) and rapid inhibition of [^{3}H]glucose incorporation into the acid-insoluble cell wall fraction of roots of dicot plants at nanomolar concentration (Sharples et al., 1998). It was also a potent inhibitor of [^{3}H]glucose incorporation into the polysaccharides of seedling roots of *Zea mays* and of *Setaria viridis*, but only relatively weakly active on *Glycine max* and *Ipomoea hederacea*.

Compound 1, as well as other thiazolidinones, has a similar syndrome of effects on plants as isoxaben. Compound 1, when used at high rates, caused severe stunting of growth and complete inhibition of seed germination. The most profound effects, however, which included swelling and splitting, were observed on roots (Sharples et al., 1998).

Although many aspects about the mode of action of compound 1 remains unknown, a common mode of action with isoxaben is suggested by the fact that the *ixr1* mutant of Arabidopsis exhibits resistance to both isoxaben and compound 1 (Scheible et al., 2001). This mode of action should differ from the mode of action of triazofenamide, since isoxaben-resistant mutants of Arabidopsis are sensitive to triazofenamide (Heim et al., 1998).

CGA 325´615

CGA 325´615 (1-cyclohexyl-5-(2,3,4,5,6-pentafluorophenoxy)-1 λ4,2,4,6-thiatriazin-3-amine) is a herbicide that inhibits the synthesis of crystalline cellulose by interfering with glucan chain crystallization. CGA 325'625 causes the accumulation of the non-crystalline β-1,4-glucan associated with CESA proteins (Peng et al., 2001).

CGA 325´615

Exposure of Arabidopsis seedlings to 10 nM CGA 325´615 for 18 h induced a radial swelling phenotype in the root tips, characteristic of phenotypes in which cellulose synthesis is inhibited. Ectopic root hair formations in this region (Kurek et al., 2002) were also induced.

The mode of action of CGA 325´615 was initially studied by Kurek et al. (2002). The authors observed that CESA subunits contain two putative zinc fingers at the N terminus. These affect rosette assembly and function through oxidative dimerization. Dimerization, which may be required as a first step for the assembly of the multimeric rosettes, is proposed to occur via a redox-regulated disulfide bond formation between at least some of the Cys that reside in the Zn domains of two CESA subunits. CGA 325´625 may affect the oxidative state of the zinc-finger domain and therefore inhibit synthesis of crystalline cellulose by interfering with the dimerization process (Kurek et al., 2002). In ccordance with this model, H_2O_2 completely reverses the effects of CGA 325´615.

Recently, it has been shown that a CGA 325'615 treatment of Arabidopsis hypocotyls had induced an internalization of CESA complexes in a microtubule-associated CESA compartment -MASC- (Crowell et al., 2009).

Accordingly to these authors, in conditions of cellulose biosynthesis inhibition, CESA complexes are recruited from the membrane into MASCs, as a way to regulate the cellulose synthesis. Interestingly, CESA complexes internalization is obtained when cell growth is impaired by osmotic stress. Furthermore, MASCs abundance is negatively correlated with hypocotyl elongation (Crowell et al., 2009). The door is open to speculate on the relationship between CBIs, cellulose biosynthesis and CESA complexes dynamic.

AE F150944

The aminotriazine AE F150944 (N2-(1-ethyl-3-phenylpropyl)-6-(1-fluoro-1-methylethyl)-1,3,5-triazine-2,4-diamine) is structurally distinct from other CBIs. AE F150944 effectively inhibited [^{14}C]glucose incorporation into crystalline cellulose with I_{50} values of 16.7·nM, 3.67·nM and 0.37·nM during primary wall synthesis in suspension cultures of the monocot *Sorghum halepense*; and secondary and primary wall synthesis in cultured cells of the dicot *Zinnia elegans*, respectively (Kiedaisch et al., 2003). Arabidopsis plants sprayed with this herbicide had accumulated several sugars, as it could be predicted from a CBI (Trenkamp et al., 2009).

AE F150944

AE F150944 specifically inhibits crystalline cellulose synthesis only in organisms that synthesize cellulose via rosettes. Although it is believed that the effect of AE F150944 is due to the destabilisation of rosettes (Kiedaisch et al., 2003), its molecular target remains to be identified. Recently, it was observed that AE F150944 mimicks several effects of an isoxaben treatment by affecting the delivery of CESA complexes to the plasma membrane, that is associated with microtubule-tethered compartments (Gutierrez et al., 2009).

Chapter 3

COMPOUNDS PROPOSED TO INDIRECTLY INHIBIT CELLULOSE BIOSYNTHESIS

Some compounds traditionally classified in other categories, such as growth retardants or auxinic herbicides, have also been shown to inhibit cellulose biosynthesis. The latter is true of quinclorac, an auxinic herbicide, and ancymidol, an inhibitor of gibberellin biosynthesis. Additionally, some antimicrotubule agents affect cellulose biosynthesis as a secondary effect.

QUINCLORAC

The quinolinecarboxylic acid quinclorac (3,7-dichloro-8-quinoline-carboxylic acid) is a highly selective and synthetic herbicide used in rice and has been developed for application in turfgrass areas, spring wheat, and chemical fallow (Grossmann, 1998). With excellent crop safety, this herbicide effectively controls dicot and monocot weeds; including grass species of *Echinochloa, Digitaria,* and *Setaria*. Quinclorac controls a broad range of about the 50 *Echinochloa* species and subspecies in existance (Lopez-Martinez et al., 1997). Quinclorac inhibits the growth of barnyard grass; particularly of the shoot, which showed progressive chlorosis, followed by wilting and necrosis (Grossmann, 1998).

Quinclorac is a systemic herbicide which is readily absorbed by germinating seeds, roots, and leaves. It is translocated in the plant both acropetally and basipetally (Lopez-Martinez et al., 1996; Grossmann, 1998).

Quinclorac

There is a controversy regarding quinclorac's mode of action. Initially, by mimicking an auxin overdose, quinclorac was proposed to affect the phytohormonal system in sensitive plants (Grossmann and Schmülling, 1995; Grossmann, 1996). The compound promotes the biosynthesis of ethylene through the stimulation of the induction of 1-aminocyclopropane-1-carboxylic acid (ACC) synthase activity. Although ethylene in-and-of-itself is not the agent that directly promotes plant death, increased levels of ethylene in susceptible grasses will provoke an accumulation of abscisic acid. It is this acid which plays a major role in growth inhibition and the induction of senescence (Grossmann et al., 2001a). Abscisic acid reduces stomatal aperture and, consequently, by the declining of photosynthetic activity; causes overproduction of the H_2O_2 which contributes to the induction of cell death. Moreover, the further oxidation of ACC, catalysed by ACC oxidase, in ethylene biosynthesis leads to an accumulation of cyanide (Grossmann, 1996, 2000); which also causes root and shoot inhibition, tissue chlorosis and necrosis. The fact that no significant differences were found in uptake, translocation, or metabolism of quinclorac between resistant and sensitive plants suggested a target-site-based mechanism of selectivity.

It was later reported, however, that some symptoms characteristic of auxinic herbicides were absent in the case of quinclorac, such as cell wall acidification due to stimulated H^+-ATPase activity (Theologis, 1987) and stimulated respiration or increased RNA content (Koo et al., 1991). Subsequently, in a dose-dependent manner in maize roots, quinclorac proved to also inhibit cell wall biosynthesis. As shown by Koo et al. (1996), after only 3 hours of treatment, the herbicide inhibited the incorporation of [^{14}C]glucose into the cell wall by 33%. The inhibitory effect increased with longer treatments and higher quinclorac concentrations; and affected not only cellulose, but also glucuronoarabinoxylans and, in a minor extent, mixed-linked glucan. In contrast, dichlobenil only inhibited cellulose biosynthesis in a parallel experiment. In a later work by Koo et al. (1997), the effect quinclorac had on the cell wall was tested in both susceptible and tolerant grasses. Cell wall biosynthesis was repressed by the herbicide by 73 and 60% in susceptible grasses, and in a minor extent (36 and 20%) in tolerant counterparts. In addition, roots of tolerant grasses were sensitive to quinclorac and their shoots were extremely tolerant, suggesting a tissue-specific response. This specific response was attributed either to the existence of an additional tolerance mechanism or to a less sensitive cell wall synthesis in shoots than in roots. In any case, based on results from these two works, Koo and co-authors proposed quinclorac to be a cell wall biosynthesis inhibitor more than an auxinic herbicide.

The controversy over the primary action of quinclorac has still remained since the later results, obtained by the Grossman group, on barnyard grass and maize showed no influence of quinclorac treatment on cellulose biosynthesis (Tresch and Grossmann, 2003). However, long treatment promoted a decline of mixed-linked glucan, similar to that promoted by dichlobenil and also by potassium cyanide. Thus, the reduction of mixed-linked glucan deposition was interpreted as an indirect effect of quinclorac, through the stimulated production of cyanide.

In order to clarify quinclorac action on cell walls, bean cultured-cells were habituated to grow in lethal concentrations of this herbicide (Alonso-Simón et al., 2008). No reduction of cellulose content was found in quinclorac-habituated cells. On the other hand, quinclorac promoted some modifications on cell walls, since habituated cells showed a lower amount of pectins in their cell walls. Consequently, quinclorac was proved not to inhibit cell wall biosynthesis in these cells. Moreover, Sunohara and Matsumoto (2008), compared quinclorac effects with those promoted by the auxin 2,4-D in maize roots. Both compounds stimulated the synthesis of ethylene and a subsequent accumulation of cyanide. Cell death rate induced by quinclorac, however, was much higher and, therefore, the toxicity of quinclorac was attributed to the reactive oxygen species production induced by this herbicide. Finally, the authors suggest that the difference in the cyanide detoxification capacity of the plant is the factor which determines if cyanide or reactive oxygen species are primarily responsible for the quinclorac herbicide action. These studies conclude that quinclorac is not a cellulose biosynthesis inhibitor.

ANCYMIDOL

Ancymidol (α-cyclopropyl-α-(4-methoxyphenyl)-5-pyrimidine methyl alcohol) -also referred to in the literature as experimental compound EL 531-, is a substituted pyrimidine with potent growth regulatory activity in higher plants.

Ancymidol

At 1 µM concentration, ancymidol has been previously shown to be an effective inhibitor of plant growth in both monocots and dicots. Since it could be overcome by gibberellin applications (Coolbaugh and Hamilton, 1976), this growth inhibition appeared to be correlated with the inhibition of gibberellin biosynthesis.

A dual action of ancymidol (an anti- gibberellin effect and gibberellin–independent activity) was reported. Early studies revealed that the primary action of ancymidol was the complete conversion blockage of ent-kaurene to ent-kaurenol in the gibberelin biosynthetic pathway; reducing the gibberellin content, further decreasing growth of ancymidol-treated plants (Coolbaugh and Hamilton, 1976; Shive and Sisler, 1976).

Besides this retardant primary action, some other effects on cell wall polysaccharides have been described. For instance, ancymidol made cells short and thick with galactose-rich cell walls in pea (Tanimoto, 1987), suppressed cell wall extensibility in dwarf pea (Tanimoto, 1994), and changed the average molecular weight of cell wall pectins (Tanimoto and Huber, 1997). Nevertheless, these modifications were reversed by gibberellin treatments, thought to be due to the inhibition of gibberellin synthesis, promoted by ancymidol. A recent work, however, described a cellulose biosynthesis inhibitor effect of ancymidol on tobacco BY-2 cells at a concentration of 1 mM, not reverted by gibberellin addition (Hofmannová et al., 2008). An ancymidol application on the cells resulted in malformations and cell death, similar to those induced by dichlobenil and isoxaben. In addition, ancymidol disoriented microtubule; and made the cellulose distribution not continuous, provoking protoplasts to regenerate a sparse net of microfibrils, or not cellulose at all, when treated with ancymidol 10 and 100 µM, respectively. This effect was made reversible by washing ancymidol from the regenerating medium, but not by the addition of gibberellin.

Many aspects of the mechanism by which ancymidol inhibits cellulose synthesis still remain unknown. It was revealed to be different than that of isoxaben, and may be related with the control of cell expansion (Hofmannová et al., 2008). It was suggested that ancymidol targets the cell wall synthesis pathway at a regulatory step, where cell wall synthesis and cell expansion are coupled.

COUMARIN AND DERIVATIVES

Other kinds of chemicals have also been shown to inhibit cellulose biosynthesis. An important group is one formed by coumarin and its derivatives. Coumarin inhibited the root elongation rate in barley (Teraoka et al., 2002), which correlated to a reduction of [^{14}C]glucose incorporated in cellulose. In cotton fibers, the reduction of [^{14}C]glucose incorporation in cellulose was not accompanied by any significant inhibition on [^{14}C]glucose incorporation in callose or noncellulosic glucans (Montezinos and Delmer, 1980). Coumarin was later observed to bind to tubulin and, thus, to suppress microtubule dynamics (Madari et al., 2003). Since the orientation of glucan microfibrils and cortical microtubules were very similar or identical, the relationship between microtubules and cellulose have been repeatedly analyzed, and the alignment hypothesis proposed that the orientation of deposited cellulose is associated with underlying cortical microtubules (Heath, 1974; Baskin, 2001). After some controversy about this alignment hypothesis, Paredez et al. (2006) visualized cellulose synthase complexes, moving along tracks, apparently defined by microtubules; confirming the functional relationship between these elements.

Coumarin

Morlin (the coumarin derivative 7-ethoxy-4-methyl chromen-2-one) was discovered in a screening for compounds that inhibited cellulose biosynthesis (DeBolt et al., 2007a). Morlin stimulated morphological defects, such as right handed helical root growth, at low concentrations; and radial swelling in elongating roots and hypocotyls, at high concentrations. Morlin was shown to decrease cellulose synthesis in a dose-dependent manner, promoting a reduction in the [^{14}C]glucose incorporation into cellulose, paralleled by an increase in callose biosynthesis. Additionally, this compound provoked a disorganization of microtubules and a reduction in cytoskeletal dynamics; diminishing the velocity of both microtubule polymerization and CESA complexes in plasma membranes, and also altering the number and spatial distribution of labeled CESA complexes (DeBolt et al., 2007a). These results support the idea of a functional interaction between microtubules synthesis or orientation and cellulose deposition. Although the spatial distribution of complexes depends on the intact microtubules, the reduced velocity did not depend on such conditions. Therefore, it is likely that morlin targets microtubule associated processes. This idea is supported by two observations: first, morlin also caused cytoskeletal defects in neuronal cells, that lack CESA complexes; and, second, a treatment with oryzalin completely depleted microtubules but did not affect CESA complex velocity.

Morlin

DeBolt and co-authors (2007a) propose three hypotheses to explain morlin effects on CESA activity and microtubule dynamics and organization: i) morlin could act separately on cellulose synthesis and microtubule dynamics (the same authors suggest that this probability is low since the high specificity of morlin action does not agree with multiple protein targets), ii) morlin might target a signaling protein that coordinates microtubule and CESA activity, and iii) morlin targets a structural protein which interacts with both microtubules and the CESA complexes. In any case, the study of morlin action and its target will be a useful tool to unravel the relationship between cortical microtubules and cellulose synthesis "machinery".

Later, a chemical genetic screening for compounds that affected the cortical microtubule-cellulose microfibrils was performed (Yoneda et al., 2007). When cortical microtubule organization or cellulose microfibril deposition is inhibited, plant cells lose their anisotropy and show swelling. Thus, compounds that caused a spherical swelling phenotype (SS compounds) were analyzed. In addition to dichlobenil, two novel compounds that reduced cellulose deposition were identified: SS14 and SS18. SS14 presents the same substructure as morlin or coumarin, but without affecting the cortical microtubule orientation at all. SS18 is a novel compound that might inhibit cellulose synthesis, directly or indirectly, by affecting substrate synthesis or transport (Yoneda et al., 2007).

COBTORIN

In the same screening performed by Yoneda et al. (2007) another compound, named SS17, later denominated as cobtorin (4-[(2-chlorophenyl)-methoxy]-1-nitrobenzene); also revealed to induce the spherical swelling phenotype on tobacco BY-2 cells. It did this by perturbing the cellulose microfibril patterning, but without neither disrupting cortical microtubules nor reducing cellulose microfibril content (see also review of Toth and Hoorn, 2010). The authors suggested that cobtorin could attack the putative linkage mechanisms between the cellulose synthase and the cortical mechanisms. The fact that cobtorin was active at a very low concentration probably indicated that the induction of the spherical swelling phenotype was not a side effect, and that cobtorin has a high binding affinity for the putative target.

Based on the following findings, a pectin mediated mechanism for cobtorin was recently proposed (Yoneda et al, 2010): i) overexpression of pectin methylesterase and polygalacturonase suppressed the cobtorin-induced cell swelling phenotype and the perturbations of cellulose microfibrils; and ii) treatment with polygalacturonase restored the deposition of cellulose microfibrils in the direction parallel to the cortical microtubules. Furthermore, it was also previously suggested that the deposition of cellulose microfibrils and their parallel orientation to the cortical microtubules was affected by distribution and methylation properties of pectin (Chanliaud and Gidley, 1999; Vignon et al., 2004; Zykwinska et al., 2005). Cobtorin would induce an increase in the methylation ratio of pectin. This could decrease the cross-linking of homogalacturonans through calcium ions and would provoke a weakening of binding between the cellulose microfibrils and pectins. The increase in the methylatyion ratio of pectin could also decreases the degree of binding of cellulose microfibrils with pectin through the side chains which consist of neutral sugars.

Cobtorin

TRIAZIFLAM

Triaziflam (N-[2-(3,5-dimethylphenoxy)-1-methylethyl]-6-(1-fluoro-1-methylethyl)-1,3,5-triazine-2,4-diamine) is an alkylazine that affects cellulose biosynthesis, amongst other cellular processes. Although the herbicidal mode of action of triaziflam was investigated (Grossmann et al., 2001b), little is known about the mode of action of alkylazines. Triaziflam has two enantiomers. The *(S)*-enantiomer, with efficacies similar to that of the herbicide atrazine, preferentially inhibited photosystem II electron transport and algae growth. In contrast, the *(R)*-enantiomer was a potent inhibitor of growth in cleaver cell suspensions and cress seedlings in the dark. At concentrations below 1 µM, the (R)-enantiomer inhibited shoot and root elongation of cress and maize seedlings The (R)-enantiomer disrupted mitosis by inhibiting microtubule formation and cellulose synthesis. Consequently, a treatment with (R)-triaziflam leads to isodiametric cell growth and to root swelling.

The analysis of these kinds of chemicals, such as triaziflam, may be useful to deepen the understanding of the cellulose synthesis mechanism and its relationships with many other cellular processes.

Triaziflam

INDAZIFLAM

Indaziflam (N-[(1R,2S)-2, 3-dihydro-2, 6-dimethyl-1-H-inden-1-yl]-6-[(1RS)-1 fluoroethyl]- 1, 3, 5-triazine-2, 4-diamine) is another kind of alkylazine, and is used as a selective herbicide that provides a long-lasting action for pre-emergence and post-emergence (when is formulated with 2,4-D, dicamba, mecoprop, and penoxsulam) control of a broad spectrum of annual grasses and broadleaf weeds. It was registered by Bayer as a cellulose biosynthesis inhibitor in 2010 (Pesticide fact sheet). Indaziflam has been described as a cellulose inhibitor that blocks cell wall formation on meristematic cells (Parrish et al., 2010).

Indaziflam

MBTU

Two enantiomers of MBTU (1-a-methylbenzyl-3-p-tolylurea) have been described. A putative mode of action for *R*-MBTU has been inferred from gene expression profiles of different clones. MBTU disrupts nitrogen metabolism, amino acid biosynthesis, cellulose synthesis, and the cell cycle. Furthermore, it represses the transcription of various genes, collectively resulting in root growth retardation. *R*-MBTU down regulates mRNA levels of endo-1,4-β-glucanase CEL1, resulting in unsuitable cell wall expansion, followed by root growth inhibition (Kojima et al., 2009, 2010).

R-MBTU

OXAZICLOMEFONE

Oxaziclomefone (3-(1-(3,5-dichlorophenyl)- 1-methylethyl)-3,4-dihydro-6-methyl-5-phenyl-2H-1,3-oxazin-4-one) is a herbicide which, at low rates, controls barnyardgrass, sedges and certain broad-leaved weeds in both paddy rice and annual grass in turf (Suzuki et al., 2003).

Oxaziclomefone inhibits the growth of gramineous plants more effectively than most other dicots (Jikihara et al., 1997). In maize cell cultures, the inhibition of growth (I_{50} of approximately 5 nM) was reported to be due to a decreased cell expansion (O'Looney and Fry, 2005a).

Oxaziclomefone

A treatment with oxaziclomefone had caused a decrease in the wall's ability to expand. The effect of oxaziclomefone was not mediated by affecting ^{14}C-incorporation from D-[U-^{14}C]glucose into the major sugar polysaccharides of the cell wall, or by the ^{14}C-incorporation from trans-[U-^{14}C]cinnamate into wall-bound ferulate or its oxidative coupling products (O'Looney and Fry 2005b). Oxaziclomefone, therefore, is not considered properly as a CBI.

Chapter 4

CBIS-RELATED MUTANTS

A relatively large number of mutants with a reduced content of cellulose has been identified in mutant screens for: i) morphological or anatomical changes, such as root or hypocotyl swelling (Baskin et al., 1992), reduced elongation of hypocotyl (Nicol et al., 1998; Desnos et al., 1996), low birefringence of the trichomes (Potikha and Delmer, 1995; Bischoff et al., 2010), altered vascular morphology (Turner and Somerville, 1997) and embryo lethality (Nickle and Meinke, 1998); or ii) CBIs tolerance such as isoxaben (Heim et al., 1989, 1990) or thiazolidinones (Scheible et al., 2001). This classic mutational analysis, from phenotypes to genes (direct genetics), presents two potential problems: due to gene redundancy, the mutations often do not cause visibly abnormal phenotypes; and in other cases, the mutations are usually lethal when the affected genes are keys to the concerned process. Therefore, it has resorted to other choices, such as: the analysis of the composition in neutral sugars by gas chromatography and mass spectroscopy (Reiter et al., 1997), tracking of FTIR spectroscopy alterations in the cell wall (Chen et al., 1998; Mouille et al., 2003; Robert et al., 2004), or the use of reverse genetics.

Table 1. Assorted data about selected CBIs

Common name	Chemical name	I_{50} bean calluses[a]	I_{50} Arabidopsis root growth	References
Dichlobenil	2,6-dichlorobenzonitrile	0.5 µM	0.4 µM[b]	Delmer, 1987; Delmer et al., 1987
Isoxaben	N-[3-(1-ethyl-1-methylpropyl)-5-isoxazolyl]-2,6-dimethoxybenzamide	3 nM	4.5 nM[c]; 1nM[d]; 1.5 nM[e]	Huggenberger et al., 1982; Heim et al., 1990
Thaxtomin A	(A 4-nitroindol-3-yl containing 2,5-dioxopiperazine)	0.6 nM	25-50 nM[f]	King et al., 1992; Fry and Loria, 2002
Flupoxam	1-[4-chloro-3-[(2,2,3,3,3-pentafluoropropoxymethyl) phenyl]-5-phenyl-1H-1,2,4-triazole-3-carboximide	2 nM	(6 nM[g])	O'Keefe and Klevorn, 1991; Hoffman and Vaughn, 1996
Triazofenamide	1-(3-methyl phenyl)-5-phenyl-1H-1,2,4-3 triazole-3-carboximide	15 nM	39 nM[h]	Heim et al., 1998
Compound 1	5-tert-butyl-carbamoyloxyl-3-(3-trifluoromethyl) phenyl-4-thiazolidinone	20 µM	<3 µM[i]	Sharples et al., 1998
CGA 325'615	1-cyclohexyl-5-(2,3,4,5,6-pentafluorophenoxy)-1 λ4,2,4,6-thiatriazin-3-amine	0.5 nM	-	Peng et al., 2001
Quinclorac	3,7-dichloro-8-quinoline carboxylic acid	10 µM	8 µM[j]	Koo et al., 1996; Koo et al., 1997; Tresch and Grossmann, 2003

Common and accepted chemical names, I_{50} on fresh weight gain of bean calluses, I_{50} on Arabidopsis root growth, and selected references about CBIs. (a) García-Angulo, unpubl.; (b) Heim et al., 1998; (c) Heim et al., 1989; (d) Heim et al., 1998; (e) Desprez et al., 2002; (f) Scheible et al., 2003; (g) Hoffmann and Vaughn, 1996 (on *Lepidium sativum* root); (h) Heim et al., 1998; (i) Sharples et al., 1998; (j) Walsh et al., 2006.

Table 2. Selected cellulose-deficient mutants in Arabidopsis

Locus	Mutant alleles	Phenotype	Gene product	Reference
CesA1	rsw1-1	Temperature-sensitive, radially expanded cells, dwarf, normal division planes	Cellulose synthase catalytic subunit, CESA1	Arioli et al., 1998
	rsw1-2	Late embryonic, radially expanded cells, dwarf, normal division plane		Gillmor et al., 2002
	rsw1-20			Beeckman et al., 2002
	rsw1-10	Post-embryonic, radially expanded cells, dwarf		Fagard et al., 2000
	eli1-1; 1-2	Post-embryonic, dwarf, accumulation ectopic lignin		Caño-Delgado et al., 2003
	cev1	Increased production of ethylene and jasmonate, stunted root	Cellulose synthase catalytic subunit,	Ellis et al., 2002
CesA3	ixr1-1; 1-2	Semi-dominant, isoxaben-resistant	CESA3	Scheible et al., 2001
	than	Root swelling, stunted growth at permissive temperature, semidominant mutation, lethaly		Daras et al., 2009
CesA4	irx5	Collapsed irregular xylem walls, thinner cell wall, weak stem, adult plants slightly smaller than wild type	Cellulose synthase catalytic subunit, CESA4	Taylor et al., 2003
CesA6	prc1-1 to 1-12	Post-embryonic, stunted root and dark-grown hypocotyls, radially expanded cells, gapped cell walls, normal microfibril orientation	Cellulose synthase catalytic subunit, CESA6	Fagard et al., 2000
	ixr2-1	Semi-dominant, isoxaben-resistant		Desprez et al., 2002
CesA7	irx3	Collapsed irregular xylem walls, thinner cell walls, weak stem, adult plants slightly smaller than wild type	Cellulose synthase catalytic subunit, CESA7	Taylor et al., 1999, 2000
CesA8	irx1	Collapsed irregular xylem walls, weak stem, adult plants slightly smaller than wild type	Cellulose synthase catalytic subunit, CESA8	Turner and Somerville, 1997; Taylor et al., 2000

Table 2. Continued

Locus	Mutant alleles	Phenotype	Gene product	Reference
Kob1	kob1-1; 1-2	Dwarf, radially expanded cells, less and disorganized microfibrils in cell elongation zone in root	Type II plasma membrane protein	Pagant et al., 2002
Cob	cob1-1	Conditional root expansion, temperature-sensitive, radially expanded cells and stunted root, reduced cell expansion in root	GPI-anchored protein	Schindelman et al., 2001
Kor1	kor1-1	Post-embryonic, dwarf, radially expanded cells	Membrane-bound endo-β-1,4-glucanase	Nicol et al., 1998
	kor1-2	Late embryonic, dwarf, randomized division planes, aborted cell plates		Zuo et al., 2000
	rsw2-1 to 4	Temperature-sensitive, radially expanded cells, dwarf, randomized division planes, aborted cell plates		Lane et al., 2001
	irx2-1; 2-2	Collapsed xylem cells, no growth phenotype		Molhoj et al., 2002
Pom1	pom1-1 to 11	Conditional root expansion, stunted root, dark-grown hypocotyls, radially expanded cells	Chitinase-like protein (AtCTL1)	Hauser et al., 1995
	elp1	Ectopic deposition of lignin, incomplete cell walls in some pith cells		Zhong et al., 2002
Cyt1	cyt1-1; cyt2	Late embryonic, increased radial expansion, incomplete cell walls, excessive callus accumulation	Mannose-1-phosphatase guanylyl-transferase	Lukowitz et al., 2001
Rsw3	rsw3-1	Temperature-sensitive, radially expanded cells, dwarf, longer lag-time before appearance phenotype, then rsw1 and 2	Glucosidase II	Burn et al., 2002
Fei	fei1; fei2 fei1 fei2 doble mutant	Disrupt anisotropic expansion in roots, defective in cellulose and other cell wall polymers, ectopic lignification	Leucine-Rich Repeat Receptor Kinases	Xu et al., 2008

Some of them have a mutation in *CesA* loci, whereas others present a mutation in other genes. Isoxaben-resistant mutants are indicated in bold.

Analysis of mutants with a reduced content of cellulose has been useful in identifing the genes involved both in cellulose synthesis and cellulose deposition in the cell wall. The mutants that will be focused on are those that are resistant to CBIs, have a changed sensitivity to CBIs, and whose phenotype is similar to CBIs-treated plants (Table 2).

MUTANTS RESISTANT TO CBIS

The screening of Arabidopsis mutants that are resistant to CBIs has allowed some clarification; not only regarding the mode of action of these compounds, but also regarding the primary cell wall composition and its organization of the cellulose biosynthesis "machinery".

Isoxaben resistant mutants (*ixr1-1* and *ixr1-2*) were the first resistant mutants identified for CBIs (Heim et al., 1989, 1990). *Ixr1* encodes the cellulose synthase catalytic subunit CESA3. Semidominant mutations, in a highly conserved region of the CESA3 subunit near the carboxyl terminus, that is well separated from the proposed active site; at the *ixr1-1* and *ixr1-2* loci, contain point mutations that replace glycine 998 with aspartic acid *(ixr1-1)* and threonine 942 with isoleucine *(ixr1-2)* (Scheible et al., 2001). Although the *Ixr1* gene is expressed in the same cells as the structurally related *rsw1* (*CesA1*) cellulose synthase gene, these two *CesA* genes are not functionally redundant (Scheible et al., 2001). Since resistant cell lines show no alterations in uptake or detoxification of the herbicide, the *ixr* mutations appear to directly affect the herbicide target (Heim et al., 1991). Mutation at *ixr-1* confers the resistance to the thiazolidinone compound 1. Since the thiazolidinone compound 1 has a similar syndrome of effects on plants as isoxaben, it was suggested that there is the same mode of action for both herbicides (Scheible et al., 2001).

Another locus was later identified in the isoxaben resistant mutant (*ixr2-1*) (Desprez et al., 2002). *Ixr2* encodes the cellulose synthase subunit CESA6. The *ixr2-1* carries a point mutation toward the highly conserved C terminus of the CESA6, causing an aminoacid change. Initially, it was thought that the presence of these two isoxaben-resistant loci (*Ixr1* and *Ixr2*) could be due to CESA3 and CESA6 being redundant. Mutants with lost function exclusively in CESA6, like *procuste1* (*prc1*, Fagard et al., 2000); however, did not restore the phenotype with the presence of CESA3. These observations suggested that CESA6 and CESA3 are activated by forming a protein complex; and the inhibitor, directly or indirectly, recognizes the place of partnership between the two subunits (see Figure 1).

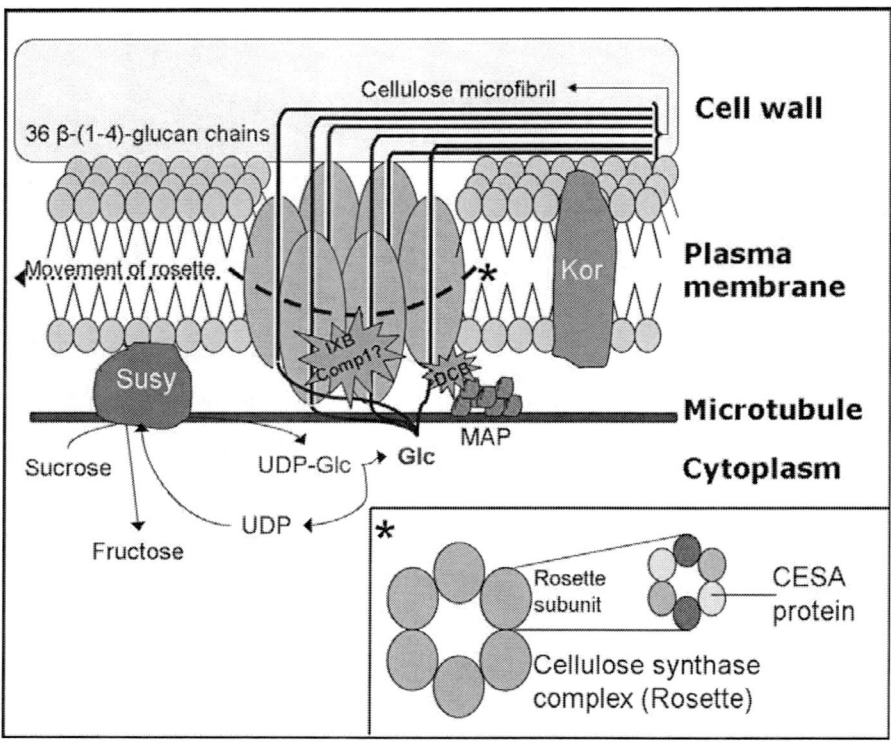

Figure 1. Model of a cellulose synthase complex (rosette). Each subunit (represented as a lobe) contains six CESA proteins, corresponding to three different CESA types (highlighted in the insert by different grey saturation). Other proteins seem to act in the cellulose biosynthesis process, such as sucrose synthase (Susy), KORRIGAN (Kor), a microtubule associated protein (MAP), and other ones, not represented here (see the text). Putative targets for some CBIs have been highlighted in boxes (IXB: isoxaben; Comp1: compound 1; DCB: dichlobenil).

Isoxaben-resistant mutants CESA3$^{ixr1-1,\ 1-2}$ and CESA6$^{ixr\ 2-1}$, along with other cell wall mutants, have contributed to demonstrate that the complex involved in the cellulose synthesis in primary cell walls actually contains three CESA catalytic subunits. Through immunolocalization analysis, co-immunoprecipitation and green fluorescent protein (GFP) gene fused expression, it was discovered that two positions of CESA complexes have been invariably occupied by CESA1 and CESA3. Additionally, in a third position at least three isoforms, CESA2, CESA5, and CESA9, may compete with CESA6; according to the tissue and/or the cell development stage (Robert et al., 2004; Desprez et al., 2007; Persson et al., 2007; Wang et al., 2008). Partial redundancy between CESA2, CESA5, CESA6 and CESA9 could explain the lower isoxaben resistance showed by CESA6^{ixr2} mutants, as compared to CESA3^{ixr1} (Desprez et al., 2007).

Other CBIs mutants appear to have different mechanisms that could be involved in the uptake and/or detoxification of the inhibitor, rather than in altered target sites. Such is the case of *txr1*, an Arabidopsis thaxtomin A-resistant mutant that presents a decrease in the rate of inhibitor uptake. The mutated gene has been cloned and encodes a highly conserved, constitutively expressed protein in eukaryotic organisms. This gene could possibly be involved in the regulation of a transport mechanism (Scheible et al., 2003).

A putative Arabidopsis mutant, resistant to dichlobenil has been reported, by using MES mutagenesis (Heim et al., 1998). This mutant (DH75) was four times more resistant to dichlobenil than the wild type. Furthermore, it did not show a cross-tolerance to either isoxaben or triazofenamide. Although the DH75 resistance-mechanism remained unravelled, an alteration in the herbicide metabolism has been pointed as its putative cause (Sabba and Vaughn, 1999).

MUTANTS WITH CHANGED SENSITIVITY TO CBIS

Currently, there are no mutants that are known to be resistant to other CBIs. Different mutants have been identified; however, with changes in sensitivity to CBIs. A mutant of Arabidopsis, named *css1* (changed sensitivity to cellulose synthesis inhibitors), grows at rates lower than the control, but it is less sensitive to isoxaben and dichlobenil (Nakagawa and Sakurai, 2006). Phenotypic analysis of this mutant, during the early developmental stages, shows changes in several metabolic processes, including: amino acid synthesis; triacylglycerides degradation; and in polysaccharides synthesis, such as cellulose or starch. The *css1* mutation seems to affect a mitochondrial protein (At-nMat1a) It was proposed that the defect in the function of mitochondria could influence cellulose synthesis through sucrose synthase, an enzyme with an important function in the connection between mitochondria and cellulose synthesis.

It has also been observed, plants with loss-of-function mutations in the genes required for cell wall synthesis (*CesA2, Cobra, Korrigan* and *CesA6*, see Somerville, 2006; *fei1 fei2* doble mutant, Xu et al., 2008, Steiwand and Kieber, 2010), were hypersensitive to CBIs such as isoxaben and dichlobenil (DeBolt et al., 2007b).

SOME MUTANTS SHOW SIMILAR PHENOTYPES TO CBIS-TREATED PLANTS

As has been previously mentioned, plants treated with such CBIs as isoxaben, dichlobenil or thaxtomin A, showed a decrease in stem growth, and root swelling. When cellulose synthesis is affected, cells expand radially and accumulate callose, lignin and other phenolic compounds (Desprez et al., 2002, Bischoff et al., 2009). This phenotype is shared by many mutants with defective hormone synthesis or signaling pathways, alterations in endocytosis processes (Collings et al., 2008), and reduced cellulose amount (Lukowitz et al., 2001). Moreover, in control plants, the application of isoxaben or dichlobenil had resulted in an incomplete cell wall formation (Desprez et al., 2002), an effect similarily observed in mutants that are deficient in cellulose (Arioli et al., 1998; Fagard et al., 2000; Lane et al., 2001). Furthermore, etiolated Arabidopsis seedlings treated with either thaxtomin A, dichlobenil, isoxaben, or flupoxam display a Fourier-transform infrared (FTIR) spectral phenotype, most closely related to those of Arabidopsis $CESA1^{rsw1}$ mutant seedlings (Scheible et al., 2003; Robert et al., 2004).

Some mutants in catalytic subunits of the CESA complex have been isolated and identified on the basis of this phenotype (Table 2), such as a set of *radial swelling 1* (*rsw1*) mutants. The *rsw1* has a mutation in the *CesA1* gene, and consists in a change of one amino acid. This results in an accumulation of non-crystalline β-1,4-glucan; a reduction in cellulose synthesis; and morphological abnormalities, including a radial expansion when plants grow at a restrictive temperature of 31°C (Baskin et al., 1992; Arioli et al., 1998; Sugimoto et al., 2001). These results suggested that the *rsw1* gene product could be required for microfibril crystallization, but not for glucan biosynthesis. Recent evidences suggest that subunits aggregation in the CESA complex changes in extracts, depending on the temperature at which CESA1^{rsw1} mutants grow. When seedlings grow at the permissive temperature; CESA proteins co-precipitate in an apparently normal CESA complex of 840 kDa; while CESA proteins remain in isolation and precipitate independently, when seedlings grow at restrictive temperature (Wang et al., 2008).

Defects in cellulose synthesis can cause ectopic lignification phenotypes and up-regulation of defense response genes. This is seen in the inhibition of cellulose synthesis in Arabidopsis by chemical inhibitors, such as either isoxaben (Caño-Delgado et al., 2003) or thaxtomin A (Bischoff et al., 2009); or in mutants with reduced content of cellulose, such as in: *rsw1, korrigan, eli1, than* and *ixr1* mutants. The *eli1* has a mutation that affects a highly conserved domain of *CesA3* gene (CesA3$^{eli1-1,\ eli1-2}$) (Caño-Delgado et al., 2000), and it causes a half reduction in the incorporation of glucose into the acid insoluble fraction of their cell walls (Caño-Delgado et al., 2003). The low level of cellulose in *eli1* is consistent with the cell shape and growth reduction in other mutants, such as CesA1^{rsw1} at the restrictive temperature (Arioli et al., 1998), and other mutants that have affected CESA in the primary cell wall as *procuste* with altered CESA6 protein (Fagard et al., 2000). This is also true of *korrigan*, which is defective in a plasma membrane associated β-1,4 endoglucanase (Nicol et al., 1998). CesA3^{ixr1-1} mutants do not show a reduction in cellulose or ectopic lignification after a treatment with isoxaben. Conversely, the control plants showed not only reduction in the synthesis of cellulose, but also revealed ectopic lignification in all plant organs, even in the most apical cells of the root that are normally actively dividing (Desprez *et al.*, 2002; Caño-Delgado et al., 2003). *Thanatos (than)* is a missense muntantion in a CesA3 with gen-dose-dependece; in which the cellulose content is reduced by one-fifth in a heterozygous plant. *Than* mutants have roots with both radial swelling and ectopic lignification. Compared with the CesA3^{ixr} mutants, *than* heterozygotes plants are not resistant to isoxaben; showing a wild-type phenotypic response (Daras et al., 2009). Many other mutants that exhibit a reduction in their cellulose amount also present an ectopic lignification (Table 2).

To date, no mutants with defects in cellulose synthesis in type II primary cell walls have been reported. The *elo* mutants, however, are a class of barley dwarfs. They have initially been described as being impaired in cell expansion (Chandler and Robertson, 1999), and have features characteristic to those similar to cellulose synthesis type I cell wall mutants. As compared with the wild-type, all *elo* mutants show radial swelling of leaf epidermal and root cortical cells and lower levels of cellulose in their leaves (Lewis et al., 2009). Although *Elo* genes have not been identified, the study of these mutants could be an excellent tool in understanding the cell expansion in plants with type II cell walls.

Chapter 5

CBIs AS TOOLS TO RESEARCH THE STRUCTURAL PLASTICITY OF CELL WALLS

By modifying their cell wall composition and structure, the ability of plant cells to tolerate induced stresses has been demonstrated in several works (Iraki et al., 1989; Shedletzky et al., 1992; Encina et al., 2001, 2002; Mélida et al., 2009). Based on the latter, CBIs are valuable tools for the analysis of cell wall structure and biogenesis (Seifert and Blaukoft, 2010). These herbicides make it possible to analyze the connections between the partially independent networks which make up the primary cell wall, and the high plasticity of this structure to accommodate to unfavourable conditions.

HABITUATION TO CBIs

Although CBIs are highly specific and potent herbicides, cell cultures of several species have been habituated to grow in the presence of CBIs by incremental exposure over many culturing cycles. Regarding the entire plant, utilizing the cell cultures is very advantageous for three reasons: i) the possibility to have lots of cells in a reduced place at the same time makes it is easy to control and manipulate different conditions; ii) the high homogenous cell response; and iii) the ability to select cell lines with specific features. The habituation of cell cultures to CBIs frequently reflects the ability of cells to survive with a modified cell wall. Therefore, this is a valuable experimental technique for gaining an insight into the plasticity of plant cell wall composition and structure. Several cell cultures have been successfully habituated to CBIs, such as: dichlobenil, isoxaben, quinclorac and thaxtomin A. Furthermore, habituated cultures usually display some common features: slower growing rates; irregularly shaped cells; a trend to grow in clumps (when suspension cultured); and cell walls with reduced cellulose contents, compensated with other cell wall components.

HABITUATION TO DICHLOBENIL

Two types of primary cell walls, having different structure and composition exist in higher plants: type I cell walls are found in dicots, gymnosperms and most monocots; and type II walls are found in graminaceous plants, along with the other commelinoid monocots (Carpita and Gibeaut, 1993; Carpita, 1996). Although the basic mechanism of habituation is common (a replacement of the cellulose network for other cell wall components), the details of the process actually depend on the type of cell wall. It has been demonstrated, therefore, that cells habituate to dichlobenil by using different strategies.

Most dichlobenil-habituated cultures belong to type I cell wall species, such as: tomato (Shedletzky et al., 1990), tobacco (Shedletzky et al., 1992; Wells et al., 1994; Nakagawa and Sakurai, 1998, 2001; Sabba et al., 1999), and bean (Encina et al., 2001, 2002; Alonso-Simón et al., 2004; García-Angulo et al., 2006). In dichlobenil-habituated type I cell walls exists a marked decrease in the amount of cellulose and hemicelluloses, whereas the quantity of esterified and unesterified pectins is increased. Moreover, in dichlobenil-habituated BY-2 tobacco cells, pectins have been reported to be cross-linked with extensins to form the main cell wall network (Sabba et al., 1999). In tomato cell cultures habituated to dichlobenil, pectins were cross-linked in the wall via phenolic-ester and/or phenolic ether linkages (Shedletzky et al., 1990, 1992). There are other modifications associated with dichlobenil habituation, such as: the presence of a non-crystalline β-1,4-glucan tightly bound to cellulose, the accumulation of pectin-enriched cell wall appositions, a putative increase in the extent of pectin–xyloglucan cross-linking and in xyloglucan endotransglucosylase activity, reduced levels of arabinogalactan proteins and changes in the levels of extensin (Shedletzky et al., 1992; Encina et al., 2002; García-Angulo et al., 2006, Alonso-Simón et al., 2007), and modifications in xyloglucan composition (Alonso-Simón et al., 2010). At least in bean cells, the type and the extent of cell wall modifications had depended both on the concentration of the inhibitor in the culture medium and on the time that the cells had been present at a given concentration of the inhibitor (Alonso-Simón et al., 2004).

To date, barley (Shedletzky et al., 1992) and maize (Mélida et al., 2009, 2010a, 2010b, 2011) are the only type II cell wall species reported to have been habituated to dichlobenil. These dichlobenil-habituated type II cell walls also displayed a modified architecture: they contained considerably reduced levels of cellulose in the cell wall, but they effectively compensated for mechanisms (parallel to modifications), quite different to those observed in dicots, and slightly different between both species. Whereas barley-habituated cultures implicated a higher proportion of mixed-linked glucan; maize-habituated cultures had a more extensive and phenolic cross-linked network of arabinoxylans, without necessitating mixed-linked glucan or other polymer enhancements. As a consequence of this modified architecture, walls from dichlobenil-habituated maize cells showed a reduction in their swelling capacity and an increase, both in pore size and in resistance to polysaccharide hydrolytic enzymes. From a molecular perspective, the application of dichlobenil to maize cell cultures disrupts the cellulose biosynthesis "machinery", affecting to the expression of several *CesA* genes. Of these, *ZmCesA5* and *ZmCesA7* seem to play a major role in habituating cells to grow in the presence of dichlobenil (Mélida et al., 2010b). Additionally, and in concordance with the increased arabinoxylans feruloylation observed, these cultures have also altered the expression of the genes involved in the synthesis of phenolic compounds. A proteomic analysis revealed that habituation to dichlobenil is linked to modifications in several metabolic pathways (Mélida et al., 2010b, 2011).

An alternative solution to tolerate dichlobenil is to control of the putative oxidative damage caused by this compound. In dichlobenil-habituated bean cells, habituation is linked to a constitutive increase in the antioxidant capacity, where guaiacol peroxidase plays a major role (García-Angulo et al., 2009a). The enhanced guaiacol peroxidase activity is stable and this could explain why bean habituated cells, cultured in a medium lacking dichlobenil for a long time (dehabituated cells), retain a high tolerance to this compound (García-Angulo et al., 2009a, 2009b). However, since there is not such a significant increase in the antioxidant capacity in habituated maize cells, this indicates that the habituation process does not rely on antioxidant strategies. Therefore, the mechanisms of habituation in cells with type II cell walls appear to be solely related to cell wall modifications.

HABITUATION TO ISOXABEN

Isoxaben-habituated cultures seem to have a more heterogeneous habituation mechanism than the dichlobenil cultures. Former isoxaben-habituated cultures had the same cellulose-xyloglucan proportions as the non-habituated ones, and the habituation seemed to be more related to changes in the herbicide target or in the detoxification than in wall modifications (Corio-Costet et al., 1991a). However, later results obtained with the French bean (Díaz-Cacho et al., 1999), tobacco (Sabba and Vaughn, 1999) and Arabidopsis cell cultures (Manfield et al., 2004), showed cell wall changes similar to those described for dichlobenil-habituated cultures. At least in Arabidopsis, isoxaben-habituation appears to be mediated neither by stress response processes, nor by functional redundancy within the CESA family (Manfield et al., 2004). Uniquely, amongst the cellulose synthase superfamily, *CslD5* is highly upregulated and might play a role in the biosynthesis of the novel walls of isoxaben-habituated cells (Bernal et al., 2007).

HABITUATION TO QUINCLORAC

Bean cells have been successfully habituated to grow in the presence of lethal concentrations of quinclorac (Alonso-Simón et al., 2008), such as it was formerly noted. Compared with non-habituated, quinclorac-habituated cells showed irregular shape and accumulated an extracellular material that was more abundant as the level of habituation increased. Cellulose content was not significantly affected by habituation. In contrast, the distribution and post-depositional modifications of pectins (mainly homogalacturonan and rhamnogalacturonan I) was affected by the habituation process. These results reflect that habituation to quinclorac is not related to cellulose biosynthesis processes.

HABITUATION TO THAXTOMIN A

Habituation to thaxtomin A has also been described in hybrid poplar cells (Brochu et al., 2010). Thaxtomin A habituated cells show changes in the composition of cell walls, similar to those observed in dichlobenil or isoxaben habituated cells: reduced levels of cellulose and an enrichment of pectins. The decrease in crystalline cellulose was much less substantial than that reported in bean cells habituated to isoxaben (Díaz-Cacho et al., 1999) or dichlobenil (Encina et al., 2001; 2002), however. Interestingly, thaxtomin A-habituated cells also exhibited enhanced tolerance to dichlobenil and isoxaben.

Tolerance to CBIs in hybrid poplar cells was associated with major changes in the expression of genes involved in many processes; including synthesis and modification of the cell wall, transcriptional regulation, lignin synthesis, and DNA and chromatin modifications. The reprogramming of gene expression, implicating modifications of DNA and chromatin, suggest the involvement of epigenetic changes (Brochu et al., 2010).

DEHABITUATION

Most of the cell wall changes induced during the dichlobenil-habituation had reverted to those of non-habituated cells when the cells were transferred to a medium without the inhibitor (Shedletzky et al., 1990; Encina et al., 2002; García-Angulo et al., 2006). Dichlobenil-dehabituated bean cell cultures retained several cell wall changes, such as: reduced levels of abinogalactan proteins and hydroxyproline-rich glycoproteins epitopes; altered extractability of pectins; and an accumulation of an amorphous β-1,4-glucan, even once crystalline cellulose level was restored (Encina et al., 2002; García-Angulo et al., 2006). Curiously, in bean cell cultures, these so-called dichlobenil-dehabituated cell cultures had retained their capacity to cope with lethal concentrations of dichlobenil. They were 40 times more tolerant to dichlobenil than non-tolerant cells (Encina et al., 2002) and also revealed a cross tolerance towards isoxaben (Encina unpubl.). As it was reported above, thaxtomin A-dehabituated poplar cells also retained tolerance to thaxtomin A and were cross tolerant to dichlobenil and isoxaben (Brochu et al., 2010). Nevertheless, dehabituated poplar cells still had reduced cellulose content and were enriched in pectins, even when cultured for more than 18 months in the absence of thaxtomin A.

It was discovered, in an attempt to explain the dichlobenil resistance of dehabituated bean cells, that these cells had constitutively increased peroxidase activity, indicating a relationship between habituation to dichlobenil and a high antioxidant capacity (García-Angulo et al., 2009a). Accordingly, the lack of an antioxidant strategy in the habituation to dichlobenil of maize cells would also explicate that the maize dehabituated cells and non-habituated cells do not differ in dichlobenil sensitivity (Mélida et al., 2010b).

Chapter 6

NEW PERSPECTIVES IN CBIS USES

CBIs represent a promising field of experimentation in order to obtain novel herbicides. Active compounds raised towards cell walls are interesting molecules to be commercialized as herbicides, taking into account their assumed lack of toxicity for non-cellulosic organisms. As it has been shown in this monography, several putative targets for CBIs have recently been hypothesized. This is a result from the new data which has been obtained, and more intense research in this field is predicted to occur.

Nevertheless, a much work is to be done to unravel the exact mechanism of action for most of these CBIs. As the cellulose biosynthesis process becomes more widely known, targets for different CBIs would also be ascertained. Reciprocally, the use of CBIs will have an important impact on cellulose biosynthesis studies.

Currently, there is a growing interest in obtaining cell walls with a modified structure, directed to change the quality and/or quantity of their components. This would allow for further study in how they could be applied to various fields, such as: food, feed, fibres or fuel. Some of these applications are oriented to obtain dietary fibres with improved properties and other significant polysaccharides in food processing, such as: pectins (Willats et al., 2006), feed products with better digestibility (Vogel and Jung, 2001), natural fibres that destined to textile and paper industries (Obembe et al., 2006); or lignocellulosic materials that more suitable for biofuels (Pauly and Keegstra, 2008; 2010).

Several approaches have been raised in order to obtain these modified cell walls: a search of mutants that have effects in genes, related to polysaccharide biosynthesis; a manipulation of genes implied in the modification of cell wall composition (Farrokhi et al., 2006); and a use of enzymes and other proteins which are able to act on cell wall components (Levy et al., 2002). A promising alternative to these approaches includes a use of CBIs. In fact, as it has been mentioned, the habituation of cell cultures to diverse CBIs, results in a modification of the cell wall composition and structure; with quantitative and qualitative changes in their components: CBIs-habituated cell walls often have a reduced content in cellulose; compensated by an increment in other polysaccharides, mainly pectins. These cell walls have been demonstrated to have new physicochemical properties, such as: modifications in pore size, their swelling capacity and resistance to polysaccharide hydrolytic enzymes (Shedletzky et al., 1992; Mélida et al., 2009). These materials should also prove to be interesting as a suitable source in study of the relationship between cellulose synthesis and other C-sink processes, such as phenylpropanoid synthesis (Mélida et al., 2009), or in the elucidation of putative new targets implied in cellulose biosynthesis.

As it has been repeatedly noted in this monography, CBIs have contributed to clarify the mechanism of cellulose biosynthesis and the organization of CESA complexes. In recent years, beside these contributions, CBIs have also exposed themselves as excellent tools, used to dig deeper in the basic processes of plant cell biology; such as in the relationship between cellulose biosynthesis and the cytoskeleton (Himmelspach et al., 2003; DeBolt et al., 2007b; Paredez et al., 2008; Wightman et al., 2009), the mechanism of cell wall sensing (Hamann et al., 2009), cell-wall based mechanisms of defense (Caño-Delgado et al., 2003; Bischoff et al., 2009), and the link between cell wall disruption and programmed cell death (Duval, 2005; Bischoff et al., 2009).

CONCLUSION

CBIs are significant compounds, not only as commercialized herbicides (or molecules with herbicide potential), but also as key tools to unravel the cellulose biosynthesis mechanism. In the last few years, a considerable amount of information regarding CBIs has been acquired, and has been directed as a tool towards new research targets. New CBIs be discovered, most likely in the future, resulting in new research on old and new compounds; opening new ways to comprehend the dynamics of the cell wall. Much has to be covered to find the mechanism of action of every member of this group of molecules, however. Surely, new advances in CBIs characterization will render a better understanding of molecular relations between cellulose and other polysaccharides; including their interactions, synthesis and potential modifications. Finally, this knowledge could be oriented to a better understanding of the plasticity in structure and composition of cell walls.

REFERENCES

Alonso-Simón, A; Encina, AE; García-Angulo, P; Álvarez, JM; Acebes, JL. FTIR spectroscopy monitoring of cell wall modifications during the habituation of bean (*Phaseolus vulgaris* L.) callus cultures to dichlobenil. *Plant Sci,* 2004 167, 1273-1281.

Alonso-Simón, A; García-Angulo, P; Encina, A; Acebes, JL; Álvarez, J. Habituation of bean (*Phaseolus vulgaris*) cell cultures to quinclorac and analysis of the subsequent cell wall modifications. *Ann Bot,* 2008 101, 1329-1339.

Alonso-Simón, A; García-Angulo, P; Encina, A; Álvarez, JM, Acebes, JL, Hayashi, T. Increase in XET activity in bean (*Phaseolus vulgaris* L.) cells habituated to dichlobenil. *Planta,* 2007 226, 765-771.

Alonso-Simón, A; Neumetzler, L; García-Angulo, P; Encina, AE; Acebes, JL; Alvarez, JM; Hayashi, T. Plasticity of xyloglucan composition in bean (*Phaseolus vulgaris*)-cultured cells during habituation and dehabituation to lethal concentrations of dichlobenil. *Mol Plant,* 2010 3, 603-609.

Amor, Y; Haigler, CH; Johnson, S; Wainscott, M; Delmer, DP. A membrane-associated form of sucrose synthase and its potential role in synthesis of cellulose and callose in plants. *Proc Natl Acad Sci USA,* 1995 92, 9353-9357.

Anderson, JR; Barnes, WS; Bedinger, P. 2,6-Dichlorobenzonitrile, a cellulose biosynthesis inhibitor, affects morphology and structural integrity of petunia and lily pollen tubes. *J Plant Physiol,* 2002 159, 61-67.

Apostolakos, P; Livanos, P; Nikolakopoulou, TL; Galatis, B. The role of callose in guard-cell wall differentiation and stomatal pore formation in the fern *Asplenium nidus*. *Ann Bot*, 2009 104, 1373-1387.

Arad, SM; Kolani, R; Simonberkovitch, B; Sivan, A. Inhibition by DCB of cell wall polysaccharide formation in the red microalga *Porphyridium sp.* (Rhodophyta). *Phycologia,* 1994 33, 158-162.

Arioli, T; Peng, L; Betzner, AS; Burn, J; Wittke, W; Herth, W; Camilleri, C; Plazinski, J; Birch, R; Cork, A; Glover, J; Redmon, J; Williamson, RE. Molecular analysis of cellulose biosynthesis in Arabidopsis. *Science,* 1998 279, 717-719.

Atanassov, II; Pittman, JK; Turner, SR. Elucidating the mechanisms of assembly and subunit interaction of the cellulose synthase complex of Arabidopsis secondary cell walls. *J Biol Chem,* 2009 284, 3833-3841.

Baskin, TI. On the alignment of cellulose microfibrils by cortical microtubules: a review and a model. *Protoplasma,* 2001 215, 150-171.

Baskin, TI; Betzner, AS; Hoggart, R; Cork, A; Williamson, RE. Root morphology mutants in *Arabidopsis thaliana. Aust J Plant Physiol,* 1992 19, 427-437.

Beeckman, T; Przemeck, GKH; Stamatiou, G; Lau, R; Terryn, N; De Rycke, R; Inzé, D; Berleth, T. Genetic complexity of cellulose synthase A gene function in Arabidopsis embryogenesis. *Plant Physiol,* 2002 130, 1883-1893.

Bernal, AJ; Jensen, JK; Harholt, J; Sorensen, S; Moller, I; Blaukopf, C; Johansen, B; Lotto, R; Pauly, M; Scheller, HV; Willats, WGT. Disruption of *ATCSLD5* results in reduced growth, reduced xylan and homogalacturonan synthase activity and altered xylan occurrence in Arabidopsis. *Plant J,* 2007 52, 1-12.

Bessueille, L. & Bulone, V. A survey of cellulose biosynthesis in higher plants. *Plant Biotechnol,* 2008 25, 315-322.

Beynon, KI. & Wright, AN. Persistence, penetration, and breakdown of chlorthiamid and dichlobenil herbicides in field soils of different types. *J Sci Food Agric,* 1968 19, 718-722.

Bignell, DRD; Huguet-Tapia, JC; Joshi, MV; Pettis, GS; Loria, R. What does it take to be a plant pathogen: genomic insights from Streptomyces species. *Antonie van Leeuwenhoek,* 2010 98, 179-194.

Bischoff, V; Cookson, SJ; Wu, S; Scheible, W. Thaxtomin A affects CESA-complex density, expression of cell wall genes, cell wall composition, and causes ectopic lignification in *Arabidopsis thaliana* seedlings. *J Exp Bot*, 2009 60, 955-965.

Bischoff, V; Nita, S; Neumetzler, L; Schindelasch, D; Urbain, A; Eshed, R; Persson, S; Delmer, D; Scheible, W. TRICHOME BIREFRINGENCE and its homolog AT5G01360 encode plant-specific DUF231 proteins required for cellulose biosynthesis in Arabidopsis. *Plant Physiol*, 2010 153, 590-602.

Bisgrove, SR & Kropf, DL. Asymmetric cell division in fucoid algae: a role for cortical adhesions in alignment of the mitotic apparatus. *J Cell Biol*, 2001 114, 4319-4328.

Blaschek, W; Semler, U; Franz, G. The influence of potential inhibitors on the invivo and invitro cell-wall beta-glucan biosynthesis in tobacco cells. *J Plant Physiol*, 1985 120, 457-470.

Bouchek-Mechiche, K; Gardan, L; Normand, P; Jouan, B. DNA relatedness among strains of *Streptomyces* pathogenic to potato in France: description of three new species, *S. europaeiscabiei* sp. nov. and *S. stelliscabiei* sp. nov. associated with common scab, and *S. reticuliscabiei* sp. nov. associated with netted scab. *Int J Syst Evol Microbiol*, 2000 50, 91-99.

Brittebo, EB; Eriksson, C; Feil,V; Bakke, J; Brandt, I. Toxicity of 2,6-dichlorothiobenzamide (chlorthiamid) and 2,6-dichlorobenzamide in the olfactory nasal mucosa of mice. *Toxicol Sci*, 1991 17, 92-102.

Brochu, V; Girard-Martel, M; Duval, I; Lerat, S; Grondin, G; Domingue, O; Beaulieu, C; Beaudoin, N. Habituation to thaxtomin A in hybrid poplar cell suspensions provides enhanced and durable resistance to inhibitors of cellulose synthesis. *BMC Plant Biol*, 2010 10, 272.

Brooker, MP. & Edwards, RW. Aquatic herbicides and the control of water weeds. *Water Res*, 1974 9, 1-15.

Brummell, DA. & Hall, JL. The role of cell wall synthesis in sustained auxin-induced growth. *Physiol Plant*, 1985 63, 406-412.

Bukhalid, RA. & Loria, R. Cloning and expression of a gene from *Streptomyces scabies* encoding a putative pathogenicity factor. *J Bacteriol*, 1997 179, 7776-7783.

Burn, JE; Hurley, UA; Birch, RJ; Arioli, T; Cork, A; Williamson, RE. The cellulose-deficient Arabidopsis mutant rsw3 is defective in a gene encoding a putative glucosidase II, an enzyme processing N-glycans during ER quality control. *Plant J,* 2002 32, 949-960.

Buron, MI. & García-Herdugo, G. Experimental analysis of cytokinesis: comparison of inhibition induced by 2,6-dichlorobenzonitrile and caffeine. *Protoplasma,* 1983 118, 192-198.

Burton, RA; Gibeaut, DM; Bacic, A; Findlay, K; Roberts, K; Hamilton, A; Baulcombe, DC; Fincher, GB. Virus-induced silencing of a plant cellulose synthase gene. *Plant Cell,* 2000 12, 691-705.

Caño-Delgado, A; Metzlaff, K; Bevan, MW. The *eli1* mutation reveals a link between cell expansion and secondary cell wall formation in *Arabidopsis thaliana. Development,* 2000 127, 3395-3405.

Caño-Delgado, A; Penfield, S; Smith, C; Catley, M; Bevan, M. Reduced cellulose synthesis invokes lignification and defense responses in *Arabidopsis thaliana. Plant J,* 2003 34, 351-362.

Carpita, NC. Structure and biogenesis of the cell walls of grasses. *Plant Physiol,* 1996 47, 445-476.

Carpita, NC. How plants make cellulose and other (1-4)-β-D-glycans. *Plant Physiol,* 2011 155, 171-184.

Carpita, NC. & Gilbeaut, DM. Structural models of primary cell walls in flowering plants: consistency of molecular structure with the physical properties of the cell walls during growth. *Plant J,* 1993 3, 1-30.

Chandler, PM. & Robertson, M. Gibberellin dose-response curves and the characterization of dwarf mutants of barley. *Plant Physiol,* 1999 120, 623-632.

Chanliaud, E; Gidley, MJ. In vitro synthesis and properties of pectin *Acetobacter xylinus* cellulose composites. *Plant J,* 1999 20, 25-35.

Chen, L; Carpita, NC; Reiter, WD; Wilson, RH; Jeffries, C; McCann, MC. A rapid method to screen for cell-wall mutants using discriminant analysis of Fourier transform infrared spectra. *Plant J,* 1998 16, 385-392.

Collings, DA; Gebbie, LK; Howles, PA; Hurley, UA; Birch, RJ; Cork, AH; Hocart, CH; Arioli, T; Williamson, RE. Arabidopsis dynamin-like protein DRP1A: a null mutant with widespread defects in endocytosis, cellulose synthesis, cytokinesis, and cell expansion. *J Exp Bot,* 2008 59, 361-376.

Coolbaugh, RC. & Hamilton, R. Inhibition of *ent*-Kaurene oxidation and growth by α-cyclopropyl-α-(*p*-methoxyphenyl)-5-pyrimidine methyl alcohol. *Plant Physiol,* 1976 57, 245-248.

Corio-Costet, MF; Dall'Agnese, M; Scalla, R (a). Effects of isoxaben on sensitive and tolerant plant cell cultures. I. Metabolic fate of isoxaben. *Pestic Biochem Physiol,* 1991 40, 246-254.

Corio-Costet, MF; Dall'Agnese, M; Scalla, R (b). Effects of isoxaben on sensitive and tolerant plant cell cultures. II. Cellular alterations and inhibition of the synthesis of acid insoluble cell wall material. *Pestic Biochem Physiol,* 1991 40, 255-265.

Crowell, EF; Bischoff, V; Desprez, T; Rolland, A; Stierhof, Y; Schumacher, K; Gonneau, M; Höfte, H; Vernhettes, S. Pausing of golgi bodies on microtubules regulates secretion of cellulose synthase complexes in Arabidopsis. *Plant Cell,* 2009 21, 1141-1154.

Crowell, EF; Gonneau, M; Stierhof, Y-D; Höfte, H; Vernhettes, S. Regulated trafficking of cellulose synthases. *Curr Op Plant Biol*, 2010 13, 705-710.

Daras G; Rigas S; Penning B; Milioni D; McCann MC; Carpita NC; Fasseas C; Hatzopoulos P. The *thanatos* mutation in *Arabidopsis thaliana* cellulose synthase 3 (*AtCesA3*) has a dominant-negative effect on cellulose synthesis and plant growth. *New Phytol*, 2009 184, 114-126.

DeBolt, S; Gutiérrez, R; Ehrhardt, DW; Melo, CV; Ross, L; Cutler, SR; Somerville, C; Bonetta, D (a). Morlin, an inhibitor of cortical microtubule dynamics and cellulose synthase movement. *Proc Natl Acad Sci USA,* 2007 104, 5854-5859.

DeBolt, S; Gutiérrez, R; Ehrhardt, DW; Somerville, C (b). Nonmotile cellulose synthase subunits repeatedly accumulate within localized regions at the plasma membrane in Arabidopsis hypocotyl cells following 2,6-dichlorobenzonitrile treatment. *Plant Physiol,* 2007 145, 334-338.

Delmer, DP. Cellulose biosynthesis. *Annu Rev Plant Physiol,* 1987 38, 259-290.

Delmer, DP. & Amor, Y. Cellulose biosynthesis. *Plant Cell,* 1995 7, 987-1000.

Delmer, DP; Read, SM; Cooper, G. Identification of a receptor protein in cotton fibers for the herbicide 2,6-dichlorobenzonitrile. *Plant Physiol*, 1987 84, 415-420.

Desnos, T; Orbovic, V; Bellini, C; Kronenberger, J; Caboche, M; Traas, J; Höfte, H. *Procuste1* mutants identify two distinct genetic pathways controlling hypocotyl cell elongation, respectively in dark- and light-grown Arabidopsis seedlings. *Development,* 1996 122, 683-693.

Desprez, T; Juraniec, M; Crowell, EF; Jouy, H; Pochylova, Z; Parcy, F; Höfte, H; Gonneau, M; Vernhettes, S. Organization of cellulose synthase complexes involved in primary cell wall synthesis in *Arabidopsis thaliana. Proc Natl Acad Sci USA,* 2007 104, 15572-15577.

Desprez, T; Vernhettes, S; Fagard, M; Refrégier, G; Desnos, T; Aletti, E; Py, N; Pelletier, S; Höfte, H. Resistance against herbicide isoxaben and cellulose deficiency caused by distinct mutations in same cellulose synthase isoform CESA6. *Plant Physiol,* 2002 128, 482-490.

Díaz-Cacho, P; Moral, R; Encina, A; Acebes, JL; Álvarez, J. Cell wall modifications in bean (*Phaseolus vulgaris*) callus cultures tolerant to isoxaben. *Physiol Plant,* 1999 107, 54-59.

Doblin, MS; Kurek, I; Jacob-Wilk, D; Delmer, DP. Cellulose biosynthesis in plants: from genes to rosettes. *Plant Cell Physiol,* 2002 43, 1407-1420.

Durso, NA. & Vaughn, KC. The herbicidal manipulation of callose levels in cell plates radically affects cell plate structure. *Plant Physiol,* 1997 114, 87(c 351).

Duval, I. & Beaudoin, N. Transcriptional profiling in response to inhibition of cellulose synthesis by thaxtomin A and isoxaben in *Arabidopsis thaliana* suspension cells. *Plant Cell Rep,* 2009 28, 811-830.

Duval, I; Brochu, V; Simard, M; Beaulieu, C; Beaudoin, N. Thaxtomin A induces programmed cell death in *Arabidopsis thaliana* suspension-cultured cells. *Planta,* 2005 222, 820-831.

Edelmann, HG. & Fry, SC. Kinetics of integration of xyloglucan into the walls of suspension-cultured rose cells. *J Exp Bot,* 1992 43, 463-470.

Eisinger, W; Croner, LJ; Taiz, L. Ethylene- induced lateral expansion in etiolated pea stems. Kinetics, cell wall synthesis, and osmotic potential. *Plant Physiol,* 1983 73, 407-412.

Ellis, C; Karafyllidis, I; Wasternack, C; Turner, JG. The Arabidopsis mutant *cev1* links cell wall signaling to jasmonate and ethylene responses. *Plant Cell,* 2002 14, 1557-1566.

Encina, A; Moral, RM; Acebes, JL; Álvarez, JM. Characterization of cell walls in bean (*Phaseolus vulgaris* L.) callus cultures tolerant to dichlobenil. *Plant Sci,* 2001 160, 331-339.

Encina, A; Sevillano, JM; Acebes, JL; Álvarez, J. Cell wall modifications of bean (*Phaseolus vulgaris*) cell suspensions during habituation and dehabituation to dichlobenil. *Physiol Plant,* 2002 114, 182-191.

Errakhi, R; Dauphin A; Meimoun, P; Lehner, A; Reboutier, D; Vatsa, P; Briand, J; Madiona, K; Rona, JP; Barakate, M; Wendehenne, D; Beaulieu, C; Bouteau, F. An early Ca^{2+} influx is a prerequisite to thaxtomin A-induced cell death in *Arabidopsis thaliana* cells. *J Exp Bot,* 2008 59, 4259-4270.

Fagard, M; Höfte, H; Vernhettes, S. Cell wall mutants. *Plant Physiol Biochem,* 2000 38, 15-25.

Fangel, JU; Petersen, BL; Jensen, NB; Willats, WGT; Bacic, A; Egelund, J. A putative *Arabidopsis thaliana* glycosyltransferase, At4g01220, which is closely related to three plant cell wall-specific xylosyltransferases, is differentially expressed spatially and temporally. *Plant Sci,* 2011 180, 470-479.

Farrokhi, N; Burton, R; Brownfield, L; Hrmova, M; Wilson, SM; Bacic, A; Fincher, GB. Plant cell wall biosynthesis: genetic, biochemical and functional genomic approaches to the identification of key genes. *Plant Biotech J,* 2006 4, 145-167.

Fisher, DD. & Cyr, RJ. Extending the microtubule/microfibril paradigm. Cellulose synthesis is required for normal cortical microtubule alignment in elongating cells. *Plant Physiol,* 1998 116, 1043-1051.

Francey, Y; Jaquet, JP; Cairoli, S; Buchala, AJ; Meier, H. The biosynthesis of β-glucans in cotton (*Gossypium hirsutum* L.) fibres of ovules cultured in vitro. *J Plant Physiol,* 1989 134, 485-491.

Fry, BA. & Loria, R. Thaxtomin A: evidence for a plant cell wall target. *Physiol Mol Plant Pathol,* 2002 60, 1-8.

García-Angulo, P; Alonso-Simón, A; Mélida, H; Encina, A; Acebes, JL; Álvarez, JM (a). High peroxidase activity and stable changes in the cell wall are related to dichlobenil tolerance. *J Plant Physiol,* 2009 166, 1229-1240.

García-Angulo, P; Alonso-Simón, A; Mélida, H; Encina, A; Álvarez, JM; Acebes, JL (b). Habituation and dehabituation to dichlobenil: simply the equivalent of Penelope's weaving and unweaving process? *Plant Sign Behav*, 2009 4, 1069-1071.

García-Angulo, P; Willats, WGT; Encina, AE; Alonso-Simón, A; Álvarez, JM; Acebes, JL. Immunocytochemical characterization of the cell walls of bean cell suspensions during habituation and dehabituation to dichlobenil. *Physiol Plant*, 2006 127, 87-99.

Gillmor, CS; Poindexter, P; Lorieau, J; Palcic, MM; Somerville, C. α-Glucosidase I is required for cellulose biosynthesis and morphogenesis in Arabidopsis. *J Cell Biol*, 2002 156, 1003-1013.

Gopalan, S. Reversal of an immunity associated plant cell death program by the growth regulator auxin. *BMC Res Notes*, 2008 1, 126.

Greenberg JT. & Yao, N. The role and regulation of programmed cell death in plant-pathogen interactions. *Cell Microbiol*, 2004 6, 201-211.

Grossmann, K. A role for cyanide, derived from ethylene biosynthesis, in the development of stress symptoms. *Physiol Plant*, 1996 97, 772-775.

Grossmann, K. Quinclorac belongs to a new class of highly selective auxin herbicides. *Weed Sci*, 1998 46, 707-716.

Grossmann, K. Mode of action of auxin herbicides: a new ending to a long, drawn out story. *Trends Plant Sci*, 2000 5, 506-508.

Grossmann, K; Kwiatkowski, J; Tresch, S (a). Auxin herbicides induce H_2O_2 overproduction and tissue damage in cleavers (*Galium aparine* L.). *J Exp Bot*, 2001 52, 1811-1816.

Grossmann, K. & Schmülling, T. The effects of the herbicide quinclorac on shoot growth in tomato is alleviated by inhibitors of ethylene biosynthesis and by the presence of an antisense construct to the 1-aminocyclopropane-1-carboxylic acid (ACC) synthase gene in transgenic plants. *Plant Growth Regul*, 1995 16, 183-188.

Grossmann, K; Tresch, S; Plath, P (b). Triaziflam and diaminotriazine derivatives affect enantioselectively multiple herbicide target sites. *Z Naturforsch [C]*, 2001 56, 559-569.

Gu, Y; Kaplinsky, N; Bringmann, M; Cobb, A; Carroll, A; Sampathkumar, A; Baskin, T; Persson, S; Somerville, C. Identification of a novel CESA-associated protein required for cellulose biosynthesis, *PNAS*, 2010 107, 12866-12871.

Guerriero, G; Fugelstad, J; Bulone, V. What do we really know about cellulose biosynthesis in higher plants? *J Integr Plant Biol*, 2010 52, 161-175.

Günther, P. & Pestemer, W. Risk assessment for selected xenobiotics by bioassay methods with higher plants. *Environ Manag,* 1990 14, 381-388.

Gutierrez, R; Lindeboom, JJ; Paredez, AR; Emons, AM; Ehrhardt, DW. Arabidopsis cortical microtubules position cellulose synthase delivery to the plasma membrane and interact with cellulose synthase trafficking compartments. *Nat Cell Biol*, 2009 11, 797-806.

Hamann, T; Bennett, M; Mansfield, J; Somerville, C. Identification of cell-wall stress as a hexose-dependent and osmosensitive regulator of plant responses. *Plant J*, 2009 57, 1015-1026.

Harris, D; Bulone, V; Ding, SY, DeBolt, S. Tools for cellulose analysis in plant cell walls, *Plant Physiol*, 2010 153, 420-426.

Hauser, MT; Mcrikami, A; Benfey, PN. Conditional root expansion mutants of Arabidopsis. *Development* 121 1995, 1237-1252.

Hazen, SP; Hawley, RM; Davis, GL; Henrissat, B; Walton, JD. Quantitative trait loci and comparative genomics of cereal cell wall composition. *Plant Physiol*, 2003 132, 263-271.

Heath, IB. Unified hypothesis for role of membrane-bound enzyme complexes and microtubules in plant cell wall synthesis. *J Theor Biol,* 1974 48, 445-449.

Heim, DR; Larrinua, IM; Murdoch, MG; Roberts, JL. Triazofenamide is a cellulose biosynthesis inhibitor. *Pestic Biochem Physiol,* 1998 59, 163-168.

Heim, DR; Roberts, JL; Phillip, DR; Larrinua, IM. A second locus, *Ixr* B1 in *Arabidopsis thaliana* that confers resistance to the herbicide Isoxaben. *Plant Physiol,* 1990 92, 695-700.

Heim, DR; Roberts, JL; Pike, PD; Larrinua, IM. Mutation of a locus of *Arabidopsis thaliana* confers resistance to the herbicide Isoxaben. *Plant Physiol,* 1989 90, 146-150.

Heim, RD; Skomp, JR; Waldron, C; Larrinua, IM. Differential response to isoxaben of cellulose biosynthesis by wild-type and resistant strains of *Arabidopsis thaliana*. *Pestic Biochem Physiol,* 1991 39, 93-99.

Herth, W. Effects of 2,6-DCB on plasma-membrane rosettes of wheat root cells. *Naturwiss,* 1987 74, 556-557.

Himmelspach, R; Williamson, RE; Wasteneys, GO. Cellulose microfibril alignment recovers from DCB-induced disruption despite microtubule disorganization. *Plant J,* 2003 36, 565-575.

Hoffman, JC. & Vaughn, KC. Flupoxam induces classic club root morphology but is not a mitotic disrupter herbicide. *Pestic Biochem Physiol,* 1996 55, 49-53.

Hofmannová, J; Schwarzerová, K; Havelková, L; Boriková, P; Petrásek, J; Opatrny, Z. A novel, cellulose synthesis inhibitory action of ancymidol impairs plant cell expansion. *J Exp Bot,* 2008 59, 3963-3974.

Hogetsu, T; Shibaoka, H; Shimokoriyama, M. Involvement of cellulose biosynthesis in actions of gibberellin and kinetin on cell expansion, 2,6-dichlorobenzonitrile as a new cellulose-synthesis inhibitor. *Plant Cell Physiol,* 1974 15, 389-393.

Hoson, T. & Masuda, Y. Role of polysaccharide synthesis in elongation growth and cell wall loosening in intact rice coleoptiles. *Planta,* 1991 155, 467-472.

Huggenberger, F; Jennings, EA; Ryan, PJ; Burow, KW. EL-107 a new selective herbicide for use in cereals. *Weeds,* 1982 1, 47-52.

Iraki, NM; Bressan, RA; Hasegawa, PM; Carpita, NC. Alteration of the physical and chemical structure of the primary cell wall of growth-limited plant cells adapted to osmotic stress. *Plant Physiol,* 1989 91, 39-47.

Jikihara, K; Maruyama, T; Morishige, J; Suzuki, H; Ikeda, K; Takagi, A; Usui, Y; Go, A. MY-100: a new herbicide for pre- and early postemergence barnyard grass control in rice. In: *Proceedings of the British Crop Protection Conference.* 1997, British Crop Protection Council, pp 73-80.

Joshi, CP. & Mansfield, SD. The cellulose paradox - simple molecule, complex biosynthesis. *Curr Opin Plant Biol,* 2007 10, 220-226.

Kiedaisch, BM; Blanton, RL; Haigler, CH. Characterization of a novel cellulose synthesis inhibitor. *Planta,* 2003 217, 922-930.

King, RR. & Calhoun, LA. The thaxtomin phytotoxins: Sources, synthesis, biosynthesis, biotransformation and biological activity. *Phytochemistry,* 2009 70, 833-841.

King, RR; Lawrence, CH; Calhoun, LA. Chemistry of phytotoxins associated with *Streptomyces scabies*, the causal organism of potato common scab. *J Agric Food Chem*, 1992 40, 834-837.

King, RR; Lawrence, CH; Gray, JA. Herbicidal properties of the Thaxtomin group of phytotoxins. *J Agric Food Chem*, 2001 49, 2298-2301.

Koo, SJ; Kwon, YW; Cho, KY. Differences in selectivity and physiological effects of quinclorac between rice and barnyardgrass compared with 2,4-D. *Proc 13th Asian-P WSSC*, 1991 103-111.

Koo, SJ; Neal, JC; DiTomaso, JM. 3,7-dichloroquinolinecarboxylic acid inhibits cell-wall biosynthesis in maize roots. *Plant Physiol*, 1996 112, 1383-1389.

Koo, SJ; Neal, JC; DiTomaso, JM. Mechanism of action and selectivity of quinclorac in grass roots. *Pestic Biochem Physiol*, 1997 57, 44-53.

Kojima, H; Hitomi, Y; Numata, T; Tanaka, C; Imai, K; Omokawa, H. Analysis of gene expression in rice root tips treated with R-1-α-methylbenzyl-3-p-tolylurea using PCR-based suppression subtractive hybridization. *Pestic Biochem Physiol*, 2009 93, 58-64.

Kojima, H; Numata, T; Tadaki, R; Omokawa, H. PCR-based suppression subtractive hybridization analyses of enantioselective gene expression in root tips of wheat treated with optically active urea compounds. *Pestic Biochem Physiol*, 2010 98, 359-369.

Kurek, I; Kawagoe, Y; Jacob-Wilk, D; Doblin, M; Delmer, D. Dimerization of cotton fiber cellulose synthase catalytic subunits occurs via oxidation of the zinc-binding domains. *Proc Natl Acad Sci USA*, 2002 99, 11109-11114.

Lambert, DH; Loria, R. *Streptomyces scabies* sp. nov., nom. rev. *Int J Syst Bacteriol*, 1989 39, 387-392.

Lane, D; Wiedemeier, A; Peng, L; Höfte, H; Vernhettes, S; Desprez, T; Hocart, CH; Birch, RJ; Baskin, TI; Burn, JE; Arioli, T; Williamson, RE. Temperature-sensitive alleles of *RSW2* link the KORRIGAN endo-1,4-β-glucanase to cellulose synthesis and cytokinesis in Arabidopsis. *Plant Physiol*, 2001 126, 278-288.

Lawrence, CH; Clark, MC; King, RR. Induction of common scab symtoms in aseptically cultured potato tubers by the vivotoxin, thaxtomin. *Phytopathology*, 1990 80, 606-608.

Lazzaro, MD; Donohue, JM; Soodavar, FM. Disruption of cellulose synthesis by isoxaben causes tip swelling and disorganizes cortical microtubules in elongating conifer pollen tubes. *Protoplasma*, 2003 220, 201-207.

Lefebvre, A; Maizonnier, D; Gaudry, JC; Clair, D; Scalla, R. Some effects of the herbicide EL-107 on cellular growth and metabolism. *Weed Res*, 1987 27, 125-134.

Leiner, RH; Fry, BA; Carling, DE; Loria, R. Probable involvement of thaxtomin A in pathogenicity of *Streptomyces scabies* on seedlings. *Phytopathology*, 1996 86, 709-713.

Levy, I; Shani, Z; Shoseyor, O. Modification of polysaccharides and plant cell wall by endo-1,4-β-glucanase and cellulose-binding domains. *Biomol Engineer*, 2002 19, 17-30.

Lewis, D; Bacic, A; Chandler, PM; Newbigin, EJ. Aberrant cell expansion in the elongation mutants of barley. *Plant Cell Physiol*, 2009 50, 554-571.

Lopez Martinez, N., De Prado, R; De Prado, R; Brown, H; Cussans GW; Devine, MD; Duke, SO; Fernandez Quintanilla, C; Helweg, A; Labrada, RE; Landes, M; Kudsk, P; Streibig, JC. 1996. Fate of quinclorac in resistant *Echinochloa crus-galli*. *Proceedings of the second international weed control congress, Copenhagen, Denmark*, 1996 1-4, 535-540.

Lopez-Martinez, N; De Prado, R; Rademacher,W; Walter, H; Marshall,G; Schmidt, O. Differential response of *Echinochloa* species and biotypes to quinclorac, in: Resistance 97, Abstract of the International Conference at IACR-Rothamsted, Harpenden, Herts, UK, 1997.

Lukowitz, W; Nickle, TC; Meinke, DW; Last, RL; Conklin, PL; Somerville, CR. Arabidopsis *cyt1* mutants are deficient in a mannose-1-phosphate guanylyltransferase and point to a requirement of N-linked glycosylation for cellulose biosynthesis. *Proc Natl Acad Sci USA*, 2001 98, 2262-2267.

Madari, H; Panda, D; Wilson, L; Jacobs, RS. Dicoumarol: a unique microtubule stabilizing natural product that is synergistic with taxol. *Cancer Res*, 2003 63, 1214-1220.

Manfield, LW; Orfila, C; McCartney, L; Harholt, J; Bernal, AJ; Scheller, HV; Gilmartin, PM; Mikkelsen, JD; Knox, JP; Willats, WGT. Novel cell wall architecture of isoxaben-habituated Arabidopsis suspension-cultured cells: global transcript profiling and cellular analysis. *Plant J*, 2004 40, 260-275.

McIntosh, AH; Chamberlain, K; Dawson, GW. Foliar sprays against potato common scab-compounds related to 3,5-dichlorophenoxyacetic acid. *Crop Prot*, 1985 4, 473-480.

Meekes, HTHM. Inhibition and recovery of cell wall formation in root hairs of *Ceratopteris thalictroides*. *J Exp Bot*, 1986 37, 1201-1210.

Meimoun, P; Tran, D; Baz, M; Errakhi, R; Dauphin, A; Lehner, A; Briand, J; Biligui, B; Madiona, K; Beaulieu, C; et al. Two different signaling pathways for thaxtomin A-induced cell death in Arabidopsis and tobacco BY2. *Plant Signal Behav*, 2009 4, 142-144.

Mélida H; Caparrós-Ruiz, D; Álvarez, J; Acebes, JL; Encina, A. Deepening into the proteome of maize cells habituated to the cellulose biosynthesis inhibitor dichlobenil. *Plant Sign Behav*, 2011 6, 143-146.

Mélida, H; García-Angulo, P; Alonso-Simón, A; Encina, A; Álvarez, J; Acebes, JL. Novel type II cell wall architecture in dichlobenil-habituated maize calluses. *Planta*, 2009 229, 617-631.

Mélida H; García-Angulo, P; Alonso-Simón, A; Álvarez, J; Acebes, JL; Encina, A (a). The phenolic profile of maize primary cell wall changes in cellulose-deficient cell cultures. *Phytochem*, 2010 71, 1684-1689.

Mélida H; Encina, A; Álvarez, J; Acebes, JL; Caparrós-Ruiz, D (b). Unraveling the biochemical and molecular networks involved in maize cells habituation to the cellulose biosynthesis inhibitor dichlobenil. *Mol Plant*, 2010 3, 842-853.

Menne, H. & Köcher, H. HRAC classification of herbicides and resistance development. In: Krämer W, Shirmer U, editors. *Modern crop protection compounds*. Wiley-Vch; 2007, pp 5-27.

Meyer, Y. & Herth, W. Chemical inhibition of cell wall formation and cytokinesis, but not of nuclear division, in protoplasts of *Nicotiana tabacum* L. cultivated in vitro. *Planta*, 1978 142, 253-262.

Miyajima, K; Tanaka, F; Takeuchi, T; Kuninaga, S. *Streptomyces turgidiscabies* sp. nov. *Int J Syst Bacteriol*, 1998 48, 495-502.

Mizuta, S. & Brown, RM. Effects of 2,6-dichlorobenzonitrile and Tinopal LPW on the structure of the cellulose synthesizing complexes of *Vaucheria hamata*. *Protoplasma*, 1992 166, 200-207.

Molhoj, M; Pagant, S; Höfte, H. Towards understanding the role of membrane-bound endo-ß-1,4-glucanases in cellulose biosynthesis. *Plant Cell Physiol*, 2002 43, 1399-1406.

Montague, MJ. Gibberellic acid promotes growth and cell wall synthesis in Avena internodes regardless of the orientation of cell expansion. *Physiol Plant,* 1995 94, 7-18.

Montezinos, D. & Delmer, DP. Characterization of inhibitors of cellulose synthesis in cotton fibers. *Planta,* 1980 148, 305-311.

Montgomery, M; Yu, TC; Freed, VH. Kinetics of dichlonenil degradation in soil. *Weed Res,* 1972 1, 31-36.

Mouille, G; Robin, S; Lecomte, M; Pagant, S; Höfte, H. Classification and identification of Arabidopsis cell wall mutants using Fourier-Transform Infrared (FT-IR) microspectroscopy. *Plant J,* 2003 35, 393-404.

Mutwil, M; DeBolt, S; Persson, S. Cellulose synthesis: a complex complex. *Curr Opin Plant Biol,* 2008 11, 252-257.

Nakagawa, N. & Sakurai, N. Increase in the amount of celA1 protein in tobacco BY-2 cells by a cellulose biosynthesis inhibitor, 2,6-dichlorobenzonitrile. *Plant Cell Physiol,* 1998 39, 779-785.

Nakagawa, N. & Sakurai, N. Cell wall integrity controls expression of endoxyloglucan transferase in tobacco BY2 cells. *Plant Cell Physiol,* 2001 42, 240-244.

Nakagawa, N. & Sakurai, N. A mutation in At-nMat1a, which encodes a nuclear gene having high similarity to group II intron maturase, causes impaired splicing of mitochondrial NAD4 transcript and altered carbon metabolism in *Arabidopsis thaliana. Plant Cell Physiol,* 2006 47, 772-783.

Nickle TC, Meinke DW. A cytokinesis-defective mutant of Arabidopsis (cyt1) characterized by embryonic lethality, incomplete cell walls, and excessive callose accumulation. *Plant J,* 1998 15, 321-332.

Nicol, F; His, I; Jauneau, A; Vernhettes, S; Canut, H; Höfte, H. A plasma membrane-bound putative endo-1,4-β-D-glucanase is required for normal wall assembly and cell elongation in Arabidopsis. *EMBO J,* 1998 17, 5563-5576.

Nishimura, MT; Stein, M; Hou, BH; Vogel, JP; Edwards, H; Somerville, S. Loss of callose synthase results in salicylic acid-dependent disease resistance. *Science,* 2003 301, 969-972.

Obembe, OO; Jacobsen, E; Visser, RGF, Vincken, JP. Cellulose-hemicellulose networks as target for *in planta* modification of the properties of natural fibres. *Biotechnol Mol Biol Rev,* 2006 1, 76-86.

References

O'Keeffe, MG; Klevorn, TB. Flupoxam: a new preemergence and postemergence herbicide for broad-leaved weed-control in winter cereals. *Brighton Crop Prot Conf Weeds*, 1991, 63-68.

O'Looney, N. & Fry, SC (a). The novel herbicide oxaziclomefone inhibits cell expansion in maize cell cultures without affecting turgor pressure or wall acidification. *New Phytol,* 2005 168, 1-7.

O'Looney, N. & Fry, SC (b). Oxaziclomefone, a new herbicide, inhibits wall expansion in maize cell-cultures without affecting polysaccharide biosynthesis, xyloglucan transglycosylation, peroxidase action or apoplastic ascorbate oxidation. *Ann Bot,* 2005 96, 1-11.

Pagant, S; Bichet, A; Sugimoto, K; Lerouxel, O; Desprez, T; McCann, M; Lerouge, P; Vernhettes, S; Hofte, H. *KOBITO1* encodes a novel plasma membrane protein necessary for normal synthesis of cellulose during cell expansion in Arabidopsis. *Plant Cell,* 2002 14, 2001-2013.

Paredez, AR; Persson, S; Ehrhardt, DW; Somerville, CR. Genetic evidence that cellulose synthase activity influences microtubule cortical array organization. *Plant Physiol,* 2008 147, 1723-1734.

Paredez, AR; Somerville, CR; Ehrhardt, DW. Visualization of cellulose synthase demonstrates functional association with microtubules. *Science,* 2006 312, 1491-1495.

Park, DH; Kim, JS; Cho, JM, Kwon, SW; Hur, JH; Lim, CK. Characterization of *Streptomyces* species causing potato scab in Korea: Distribution, taxonomy, and pathogenicity. *Plant Pathol J,* 2003 19, 13-18.

Parrish, MD; Unland, RD; Bertges, WJ. Introduction of indaziflam for weed control in fruit, nut, and grape crops. Bayer CropScience, Research Triangle Park, NC 27709. (164). 2010.

Pauly, M. & Keegstra, K. Cell-wall carbohydrates and their modification as a resource for biofuels. *Plant J,* 2008 54, 569-568.

Pauly, M. & Keegstra, K. Plant cell wall polymers as precursors for biofuels. *Curr Op Plant Biol,* 2010 13, 304-311.

Pedersen, CT. & Sylvia, DM. Mycorrhiza: ecological implication of plant interactions. In: Mukerji, editor. *Concepts in Mycorrizhal Research,* 1996 19, pp 195-222.

Peng, L; Kawagoe, Y; Hoan, P; Delmer, D. Sitosterol-β-glucosidase as primer for cellulose synthesis in plants. *Science,* 2002 295, 147-150.

Peng, L; Xiang, F; Roberts, E; Kawagoe, Y; Greve, LC; Kreuz, K; Delmer, DP. The experimental herbicide CGA 325´615 inhibits synthesis of crystalline cellulose and causes accumulation of non-crystalline normal β-1,4-glucan associated with CesA protein. *Plant Physiol,* 2001 126, 981-992.

Persson, S; Paredez, A; Carroll, A; Palsdottir, H; Doblin, M; Poindexter, P; Khitrov, N; Auer, M; Somerville, CR. Genetic evidence for three unique components in primary cell-wall cellulose synthase complexes in Arabidopsis. *Proc Natl Acad Sci USA,* 2007 104, 15566-15571.

Potikha, T. & Delmer, DP. A mutant of *Arabidopsis thaliana* displaying altered patterns of cellulose deposition. *Plant J,* 1995 7, 453-460.

Rajangam, AS; Kumar, M; Aspeborg, H; Guerriero, G; Arvestad, L; Pansri, P; Brown, CJ-; Hober, S; Blomqvist, K; Divne, C; Ezcurra, I; Mellerowicz, E; Sundberg, B; Bulone, V; Teeri, TT. MAP20, a microtubule-associated protein in the secondary cell walls of hybrid aspen, is a target of the cellulose synthesis inhibitor 2,6-dichlorobenzonitrile. *Plant Physiol,* 2008 148, 1283-1294.

Reiter, W-; Chapple, C; Somerville, CR. Mutants of *Arabidopsis thaliana* with altered cell wall polysaccharide composition. *Plant J,* 1997 12, 335-345.

Richmond, TA. & Somerville, CR. The cellulose synthase superfamily. *Plant Physiol,* 2000 124, 495-498.

Robert, S; Mouille, G; Höfte, H. The mechanism and regulation of cellulose synthesis in primary walls: lessons from cellulose-deficient Arabidopsis mutants. *Cellulose,* 2004 11, 351-364.

Rouchaud, J; Gustin, F; Van-Himme, M; Bulcke, R; Sarrazyn, R. Soil dissipation of the herbicide isoxaben after use in cereals. *Weed Res,* 1993 33, 205-212.

Rudolph, U; Gross, H; Schnepf, E. Investigations of the turnover of the putative cellulose-synthesizing particle "rosettes" within the plasma membrane of *Funaria hygrometrica* protonema cells II. Rosette structure and the effects of cycloheximide, actinomycin D, 2,6-dichlorobenzonitrile, biofluor, heat shock, and plasmolysis. *Protoplasma,* 1989 148, 57-69.

Sabba, RP; Durso, NA; Vaughn, KC. Structural and immunocytochemical characterization of the walls of dichlobenil-habituated BY-2 tobacco cells. *Int J Plant Sci,* 1999 160, 275-290.

Sabba, RP. & Vaughn, KC. Herbicides that inhibit cellulose biosynthesis. *Weed Sci,* 1999 47, 757-763.

Samuels, AL; Giddings, TH; Staehelin, LA. Cytokinesis in tobacco BY-2 and root tip cells: a new model of cell plate formation in higher plants. *J Cell Biol,* 1995 130, 1345-1357.

Scheible, W-; Eshed, R; Richmond, T; Delmer, D; Somerville, C. Modifications of cellulose synthase confer resistance to isoxaben and thiazolidinone herbicides in Arabidopsis *Ixr1* mutants. *Proc Natl Acad Sci USA,* 2001 98, 10079-10084.

Scheible, W-; Fry, B; Kochevenko, A; Schindelasch, D; Zimmerli, L; Somerville, S; Loria, R; Somerville, CR. An Arabidopsis mutant resistant to thaxtomin A, a cellulose synthesis inhibitor from *Streptomyces* species. *Plant Cell,* 2003 15, 1781-1794.

Schindelman, G; Morikami, A; Jung, J; Baskin, TI; Carpita, NC; Derbyshire, P; McCann, MC; Benfey, PN. COBRA encodes a putative GPI-anchored protein, which is polarly localized and necessary for oriented cell expansion in Arabidopsis. *Genes Develop,* 2001 15, 1115-1127.

Schneegurt, MA; Roberts, JL; Bjelk, LA; Gerwick, BC. Postemergence activity of isoxaben. *Weed Technol,* 1994 8, 183-189.

Seifert, GJ. Nucleotide sugar interconversions and cell wall biosynthesis: how to bring the inside to the outside. *Curr Opin Plant Biol,* 2004 7, 277-284.

Seifert, GJ. & Blaukopf, C. Irritable walls: the plant extracellular matrix and signaling. *Plant Physiol,* 2010 153, 467-478.

Sharples, K; Hawkes, TR; Mitchell, G; Edwards, LS; Langford, MP; Langton, DW; Rogers, KM; Townson, JK; Wang, Y. A novel thiazolidinone herbicide is a potent inhibitor of glucose incorporation into cell wall material. *Pestic Sci,* 1998 54, 368-376.

Shedletzky, E; Shmuel, M; Delmer, DP; Lamport, DTA. Adaptation and growth of tomato cells on the herbicide 2,6-Dichlorobenzonitrile leads to production of unique cell-walls virtually lacking a cellulose-xyloglucan network. *Plant Physiol,* 1990 94, 980-987.

Shedletzky, E; Shmuel, M; Trainin, T; Kalman, S; Delmer, D. Cell-wall structure in cells adapted to growth on the cellulose-synthesis inhibitor 2,6-Dichlorobenzonitrile - A comparison between two dicotyledonous plants and a graminaceous monocot. *Plant Physiol,* 1992 100, 120-130.

Shive, JB. & Sisler, HD. Effects of Ancymidol (a growth retardant) and Triarimol (a fungicide) on growth, sterols, and gibberellins of *Phaseolus vulgaris* L. *Plant Physiol,* 1976 57, 640-644.

Somerville, C. Cellulose synthesis in higher plants. *Annu Rev Cell Dev Biol,* 2006 22, 53-78.

Steinwand, BJ. & Kieber, JJ. The role of receptor-like kinases in regulating cell wall function. *Plant Physiol,* 2010 153, 479-484.

Sugimoto, K; Williamson, R; Wasteneys, GO. Wall architecture in the cellulose-deficient *rsw1* mutant of *Arabidopsis thaliana*: microfibrils but not microtubules lose their transverse alignment before microfibrils become unrecognizable in the mitotic and elongation zones of roots. *Protoplasma,* 2001 215, 172-183.

Sunohara, Y. & Matsumoto, H. Quinclorac-induced cell death is accompanied by generation of reactive oxygen species in maize root tissue. *Phytochemistry,* 2008 69, 2312-2319.

Suzuki, H; Jikihara, K; Sonoda, M; Usui, Y. Development of a new herbicide, oxaziclomefone. *J Pest Sci,* 2003 28, 241-248.

Tanimoto, E. Gibberellin-dependent root elongation in *Lactuca sativa:* Recovery from growth retardant-suppressed elongation with thickening by low concentration of GA_3. *Plant Cell Physiol,* 1987 28, 963-973.

Tanimoto, E. Interaction of gibberellin A_3 and ancymidol in the growth and cell-wall extensibility of dwarf pea roots. *Plant Cell Physiol,* 1994 35, 1019-1028.

Tanimoto, E. & Huber, DJ. Effect of GA_3 on the molecular mass of polyuronides in the cell walls of Alaska pea roots. *Plant Cell Physiol,* 1997 38, 25-35.

Taylor, NG. Cellulose biosynthesis and deposition in higher plants *New Phytol,* 2008 178, 239-252.

Taylor, NG; Howells, RM; Huttly, AK; Vickers, K; Turner, SR. Interactions among three distinct CesA proteins essential for cellulose synthesis. *Proc Natl Acad Sci USA,* 2003 100, 1450-1455.

Taylor, NG; Laurie, S; Turner, SR. Multiple cellulose synthase catalytic subunits are required for cellulose synthesis in Arabidopsis. *Plant Cell,* 2000 12, 2529-2540.

Taylor, NG; Scheible, WR; Cutler, S; Somerville, CR; Turner, SR. The *irregular xylem3* locus of Arabidopsis encodes a cellulose synthase required for secondary cell wall synthesis. *Plant Cell*, 1999 11, 769-780.

Tegg, RS; Gill, WM; Thompson, HK; Davies, NW; Ross, JJ; Wilson, CR. Auxin-induced resistance to common scab disease of potato linked to inhibition of thaxtomin A toxicity. *Plant Dis,* 2008 92, 1321-1328.

Tegg, RS; Melian, L; Wilson, CR; Shabala, S. Plant cell growth and ion flux responses to the streptomycete phytotoxin thaxtomin A: Calcium and hydrogen flux patterns revealed by the non-invasive MIFE technique. *Plant Cell Physiol*, 2005 46, 638-648.

Theologis, A. Possible linkage between auxin regulated gene expression, H^+ secretion, and cell elongation: a hypothesis. In Cosgrove DJ, Knievel DP editors. *Physiology of cell expansion during plant growth.* American Society of Plant Physiologists, Rockville, MD, 1987 pp 133-144.

Teraoka, T; Kaneko, M; Mori, S; Yoshimura, E. Aluminum rapidly inhibits cellulose synthesis in roots of barley and wheat seedlings. *J Plant Physiol*, 2002 159, 17–23.

Timmers, J; Vernhettes, S; Desprez, T; Vincken, J; Visser, RGF; Trindade, LM. Interactions between membrane-bound cellulose synthases involved in the synthesis of the secondary cell wall. *FEBS Lett,* 2009 583, 978-982.

Toth, R. & van der Hoorn, RAL. Emerging principles in plant chemical genetics. *Trends Plant Sci*, 2010 15, 81-88.

Trenkamp, S; Eckes, P; Busch, M; Fernie, AR. Temporally resolved GC-MS-based metabolic profiling of herbicide treated plants treated reveals that changes in polar primary metabolites alone can distinguish herbicides of differing mode of action. *Metabolomics*, 2009 5, 277-291.

Tresch, S. & Grossmann, K. Quinclorac does not inhibit cellulose (cell wall) biosynthesis in sensitive barnyard grass and maize roots. *Pestic Biochem Physiol,* 2003 75, 73-78.

Turner, SR. & Somerville, CR. Collapsed xylem phenotype of Arabidopsis identifies mutants deficient in cellulose deposition in the secondary cell wall. *Plant Cell,* 1997 9, 689-701.

Umetsu, N; Satoh, S; Matsuda, K. Effects of 2,6-dichlorobenzonitrile on suspension-cultured soybean cells. *Plant Cell Physiol,* 1976 17, 1071-1073.

Vaughn, KC. Cellulose biosynthesis inhibitor herbicides. In: Böger P, Wakabayashi K, Hirai K, editors. *Herbicide classes in development.* Springer, Berlin, 2002, pp 139-150.

Vaughn, KC; Hoffman, JC; Hahn, MG; Staehelin, LA. The herbicide dichlobenil disrupts cell plate formation: immunogold characterization. *Protoplasma,* 1996 194, 117-132.

Vaughn, KC. & Turley, RB. Ultrastructural effects of cellulose biosynthesis inhibitor herbicides on developing cotton fibers. *Protoplasma,* 2001 216, 80-93.

Verloop, A. Fate of the herbicide dichlobenil in plants and soil in relation to its biological activity. *Residue Rev,* 1972 43, 55-103.

Verloop, A. & Nimmo, WB. Absorption, translocation and metabolism of dichlobenil in bean seedlings. *Weed Res,* 1969 9, 357-370.

Verloop, A. & Nimmo, WB. Transport and metabolism of dichlobenil in wheat and rice seedlings. Weed Res, 1970 10, 59-64.

Vignon, MR; Heux, L; Malainine, ME; Mahrouz; M. Arabinan-cellulose composite in *Opuntia ficus-indica* prickly pear spines. *Carbohydr Res,* 2004 339, 123-131.

Vogel, KP. & Jung, HJG. Genetic modification of herbaceous plants for feed and fuel. *Crit Rev Plant Sci,* 2001 20, 15-49.

Wakabayashi, K. & Böger, P. Phytotoxic sites of action for molecular design of modern herbicides (Part 2): Amino acid, lipid and cell wall biosynthesis, and other targets for future herbicides. *Weed Biol Manag,* 2004 4, 59-70.

Walsh, TA; Neal, R; Merlo, AO; Honma, M; Hicks, GR; Wolff, K; Matsumura, W; Davies, JP. Mutations in an auxin receptor homolog AFB5 and in SGT1b confer resistance to synthetic picolinate auxins and not to 2,4-dichlorophenoxyacetic acid or indole-3-acetic acid in Arabidopsis. *Plant Physiol,* 2006 142, 542-552.

Wang, J; Elliott, JE; Williamson, RE. Features of the primary wall CESA complex in wild type and cellulose-deficient mutants of *Arabidopsis thaliana. J Exp Bot,* 2008 59, 2627-2637.

Wang, Y; Lu, J; Mollet, J-; Gretz, MR; Hoagland, KD. Extracellular matrix assembly in diatoms (Bacillariophyceae). II. 2,6-dichlorobenzonitrile inhibition of motility and stalk production in the marine diatom *Achnanthes longipes. Plant Physiol,* 1997 113, 1071-1080.

Wells, B; McCann, MC; Shedletzky, E; Delmer, D; Roberts, K. Structural features of cell walls from tomato cells adapted to grow on the herbicide 2,6-dichlorobenzonitrile. *J Microsc,* 1994 173, 155-164.

Wightman, R; Marshall, R; Turner, SR. A cellulose synthase-containing compartment moves rapidly beneath sites of secondary wall synthesis. *Plant Cell Physiol,* 2009 50, 584-594.

Willats, WGT; Knox, P; Mikkelsen, D. Pectin: new insights into an old polymer are starting to gel. *Trends Food Sci Technol,* 2006 17, 97-104.

Wilson, CR; Luckman, GA; Tegg, RS; Yuan, ZQ; Wilson AJ; Eyles, A; Conner, AJ. Enhanced resistance to common scab of potato through somatic cell selection in cv. Iwa with the phytotoxin thaxtomin A. *Plant Pathol,* 2009 58, 137-144.

Xu, SL; Rahman, A; Baskin, TI; Kieber, JJ. Two leucine-rich repeat receptor kinases mediate signaling, linking cell wall biosynthesis and ACC synthase in *Arabidopsis. Plant Cell,* 2008 20, 3065-3079.

Yoneda, A; Higaki, T; Kutsuna, N; Kondo, Y; Osada, H; Hasezawa, S; Matsui, M. Chemical genetic screening identifies a novel inhibitor of parallel alignment of cortical microtubules and cellulose microfibrils. *Plant Cell Physiol,* 2007 48, 1393-1403.

Yoneda, A; Ito, T; Higaki, T; Kutsuna, N; Saito, T; Ishimizu, T; Osada, H; Hasezawa, S; Matsui, M; Demura, T. Cobtorin target analysis reveals that pectin functions in the deposition of cellulose microfibrils in parallel with cortical microtubules. *Plant J,* 64 2010, 657-667.

Zhong, R; Kays, SJ; Schroeder, BP; Ye, ZH. Mutation of a chitinase-like gene causes ectopic deposition of lignin, aberrant cell shapes, and overproduction of ethylene. *Plant Cell,* 2002 14, 165-179.

Zuo, J; Niu, QW; Nishizawa, N; Wu, Y; Kost, B; Chua, NH. KORRIGAN, an Arabidopsis endo-1,4-beta-glucanase, localizes to the cell plate by polarized targeting and is essential for cytokinesis. *Plant Cell,* 2000 12, 1137-1152.

Zykwinska, AW; Ralet, MCJ; Garnier, CD; Thibault, JFJ. Evidence for in vitro binding of pectin side chains to cellulose. *Plant Physiol,* 2005 139, 397-407.

INDEX

A

acetic acid, 76
acetone, 11
acetonitrile, 11
acid, 21, 25, 26, 45, 61, 67, 69, 70, 76
active site, 14, 41
adhesions, 59
aggregation, 45
algae, 7, 8, 10, 33, 59
alters, 9
amine, 21, 38
amino, 35, 43, 44
anisotropy, 15, 31
antioxidant, 49, 52
antisense, 64
arabinogalactan, 49
arrests, 13
aspartic acid, 41
assessment, 65
auxins, 18, 76

B

bioassay, 65
biological activity, 66, 76
biological processes, 5
biopolymer, vii
biosynthesis, vii, 1, 2, 3, 7, 8, 9, 11, 13, 14, 15, 17, 18, 19, 20, 23, 25, 26, 27, 28, 29, 30, 33, 34, 35, 41, 42, 45, 49, 50, 51, 53, 54, 55, 57, 58, 59, 61, 62, 63, 64, 65, 66, 67, 68, 69, 70, 71, 73, 74, 75, 76, 77
biotic, 5
birefringence, 4, 37
body weight, 6
breakdown, 58

C

caffeine, 60
calcium, 32
cancer, 6
carbohydrates, 71
carbon, 70
carboxyl, 14, 26, 38, 41, 64
carcinogen, 6
challenges, 15
chemical, 11, 19, 25, 29, 31, 34, 38, 45, 66, 75
chitinase, 77
chromatography, 37
circulation, 10
classes, 76
classification, vii, 1, 69

cloning, 14
colonization, 16
color, iv
compensation, 18
competition, 19
complexity, 58
composites, 60
composition, vii, viii, 15, 17, 19, 37, 41, 47, 48, 49, 51, 54, 55, 57, 59, 65, 72
compounds, vii, 1, 25, 28, 30, 31, 41, 53, 55, 67, 69
concordance, 49
congress, 68
conifer, 15, 68
contamination, 5
controversial, 9
cortex, 8
cotton, 9, 19, 29, 61, 63, 67, 70, 76
crop, 25, 69
crops, 12, 20, 71
crystalline, 8, 9, 10, 11, 21, 22, 23, 24, 44, 48, 51, 52, 72
crystallization, 2, 3, 9, 22, 45
culture, 16, 49
culture media, 16
culture medium, 49
cuticle, 12
cyanide, 26, 27, 28, 64
cycles, 47
cytokinesis, 60, 67, 69, 70, 77
cytoskeleton, 15, 54

D

death rate, 28
decoupling, 11
defects, 30, 46, 60
defense mechanisms, 15
deficiency, 62
degradation, 43, 70

deposition, vii, 1, 2, 4, 8, 12, 15, 27, 30, 31, 32, 40, 41, 72, 74, 75, 77
derivatives, vii, 29, 64
detachment, 12
detoxification, 14, 28, 41, 43, 50
diatoms, 76
digestibility, 53
dimerization, 22
discriminant analysis, 60
distribution, 3, 29, 30, 32, 51
drugs, 1

E

editors, 69, 75, 76
electron, 2, 8, 33
electron microscopy, 2
elongation, 13, 16, 23, 29, 33, 37, 40, 62, 66, 68, 70, 74, 75
elucidation, 54
embryogenesis, 58
enantiomers, 33, 35
encoding, 15, 59, 60
enzyme, 14, 18, 44, 60, 65
enzymes, 49, 54
ester, 48
ethylene, 15, 26, 28, 39, 62, 64, 77
eukaryotic, 43
evidence, 3, 17, 63, 71, 72
exocytosis, 3
experimental condition, 9
exposure, 47
extracellular matrix, 73
extracts, 45
extrusion, 14

F

fatty acids, 13
fiber, 9, 10, 19, 29, 61, 67, 70, 76
fish, 6

food, 53
force, 9
formation, 2, 3, 7, 9, 10, 11, 13, 14, 22, 33, 34, 44, 58, 60, 69, 73, 76
fungi, 6
fusion, 10

G

gel, 77
gene expression, 18, 35, 51, 67, 75
genes, 3, 4, 11, 15, 16, 17, 18, 19, 31, 35, 37, 40, 41, 44, 45, 46, 49, 51, 54, 59, 62, 63, 75, 77
genomics, 65
germination, 6, 21
gibberellin, 25, 28, 29, 66, 74
glucose, 2, 3, 7, 9, 13, 14, 17, 20, 21, 23, 27, 29, 30, 36, 45, 73
glucoside, 2, 9
glycans, 60
glycine, 41
glycoproteins, 52
glycosylation, 68
grass, 20, 25, 26, 27, 34, 35, 60, 66, 67, 75
growth, 2, 5, 6, 7, 12, 16, 17, 21, 23, 25, 26, 28, 30, 33, 35, 38, 39, 40, 44, 45, 58, 59, 60, 61, 64, 66, 68, 70, 73, 74, 75

H

habituation, viii, 2, 47, 48, 49, 50, 51, 52, 54, 57, 63, 64, 69
hemicellulose, 70
herbicide, 1, 5, 11, 14, 19, 21, 23, 25, 26, 27, 33, 34, 35, 41, 43, 50, 55, 61, 62, 64, 65, 66, 68, 71, 72, 73, 74, 75, 76, 77
hormone, 44
host, 16
human, 6
hybrid, 10, 51, 59, 72

hybridization, 67
hydrogen, 75
hypertrophy, 16
hypocotyl, 7, 13, 23, 37, 61, 62
hypothesis, 9, 30, 65, 75

I

identification, 63, 70
immunity, 64
immunolocalization, 42
immunoprecipitation, 43
in vitro, 2, 9, 11, 12, 63, 69, 77
in vivo, 9, 11
indirect effect, 9, 27
induction, 19, 26, 32
industries, 53
industry, 2
infection, 16
inhibition, vii, 6, 7, 9, 13, 15, 17, 18, 21, 23, 26, 28, 29, 35, 45, 60, 61, 62, 69, 75, 76
inhibitor, 6, 15, 16, 20, 21, 25, 27, 28, 29, 33, 34, 42, 43, 49, 52, 57, 61, 65, 66, 69, 70, 72, 73, 76, 77
initiation, 2, 9
injury, iv
integration, 62
integrity, 15, 57, 70
interference, 10
internalization, 4, 22
ions, 32
isolation, vii, 45
isoleucine, 41

L

landscape, 12
lead, 15
lesions, 7, 16
leucine, 77
light, 62

lignin, 15, 19, 39, 40, 44, 51, 77
loci, 14, 40, 41, 65
locus, 10, 41, 65, 75

M

machinery, vii, 11, 31, 41, 49
manipulation, 54, 62
marine diatom, 76
mass, 37
materials, 53, 54
matrix, 76
membranes, 8, 10
metabolic pathways, 49
metabolism, 5, 18, 26, 35, 43, 68, 70, 76
metabolites, 75
metabolized, 11
methylation, 32
mice, 59
microorganism, 5
mitochondria, 44
mitosis, 33
models, 60
modifications, 17, 27, 29, 48, 49, 50, 51, 54, 55, 57, 62, 63
molecules, 53, 55
morphogenesis, 64
morphological abnormalities, 44
morphology, 19, 37, 57, 58, 66
mucosa, 6, 59
mutagenesis, 43
mutant, 10, 20, 21, 37, 40, 41, 43, 44, 60, 62, 70, 72, 73, 74
mutation, 10, 14, 15, 37, 39, 41, 40, 43, 44, 45, 60, 61, 62, 70

N

necrosis, 25, 26
neuronal cells, 30
neutral, 32, 37

nitrobenzene, 32
nitrogen, 35
nucleic acid, 13
null, 60

O

organic solvents, 11
organism, 67
organs, 13, 45
osmotic stress, 23, 66
overproduction, 26, 64, 77
oxidation, 26, 61, 67, 71
oxidative damage, 49
oxygen, 28

P

parallel, 27, 32, 49, 77
pathways, 1, 15, 18, 62
pesticide, 5
phenolic compounds, 44, 49
phenotype, 13, 19, 22, 31, 32, 40, 41, 42, 44, 75
phenotypes, 22, 37, 45
phloem, 6
phosphate, 68
photosynthesis, 13
physical properties, 60
physicochemical properties, 54
pith, 40
plants, 1, 2, 7, 8, 10, 11, 12, 13, 14, 15, 16, 21, 23, 26, 28, 35, 39, 41, 44, 45, 46, 48, 57, 58, 60, 61, 62, 64, 65, 71, 73, 74, 75, 76
plasma membrane, 3, 4, 8, 10, 12, 14, 18, 24, 30, 40, 45, 61, 65, 70, 71, 72
plasmolysis, 72
plasticity, viii, 2, 47, 48, 55
point mutation, 41
polar, 75

pollen, 7, 15, 57, 68
polymer, 4, 49, 77
polymerization, 2, 8, 9, 20, 30
polymers, 40, 71
polypeptide, 10, 11
polysaccharide, 49, 54, 58, 66, 71, 72
potassium, 27
potato, 16, 59, 67, 69, 71, 75, 77
preparation, iv
principles, 75
probability, 31
protection, 69
protein synthesis, 13, 18
proteins, 2, 3, 4, 11, 14, 15, 22, 42, 45, 49, 52, 54, 59, 74
proteome, 69
protonema, 72
putative cause, 43
pyrimidine, 28, 61

Q

quality control, 60

R

radicle, 6
reactions, 6
reactive oxygen, 28, 74
recognition, 14
recovery, 69
redundancy, 37, 43, 50
regenerate, 16, 29
regeneration, 7
reinforcement, 19
reproduction, 6
residues, 2
resistance, 21, 41, 43, 49, 52, 54, 59, 65, 69, 70, 73, 75, 76, 77
respiration, 13, 27

response, 12, 15, 27, 45, 47, 50, 60, 62, 65, 68
retardation, 35
root, 5, 6, 7, 8, 9, 12, 13, 16, 18, 19, 21, 22, 26, 27, 28, 29, 30, 33, 35, 37, 38, 39, 40, 44, 45, 46, 65, 66, 67, 69, 73, 74, 75

S

safety, 25
saturation, 42
scabies, 16, 59, 67, 68
secretion, 4, 61, 75
seed, 6, 21
seedlings, 6, 13, 15, 17, 18, 19, 22, 33, 44, 45, 59, 62, 68, 75, 76
selectivity, 26, 67
senescence, 26
sensing, 54
sensitivity, 41, 43, 52
shape, 45, 51
shock, 72
shoot, 12, 25, 26, 33, 64
shoots, 27
showing, 1, 45
side chain, 32, 77
signaling pathway, 44, 69
signals, 15
solution, 49
somatic cell, 77
species, 1, 11, 13, 16, 18, 20, 25, 28, 47, 48, 49, 58, 59, 68, 71, 73, 74
spectroscopy, 37
stability, 15
starch, 43
state, 4, 22
sterols, 74
stress, 15, 18, 50, 64, 65
structural protein, 31
structure, viii, 3, 15, 19, 47, 48, 53, 54, 55, 62, 66, 69, 72, 73

substrate, 2, 31
substrates, 3
sucrose, 3, 42, 44, 57
suppression, 16, 67
suspensions, 12, 33, 59, 63, 64
swelling, 7, 12, 16, 21, 22, 30, 31, 32, 33, 37, 39, 44, 45, 46, 49, 54, 68
symptoms, 7, 12, 16, 20, 26, 64
syndrome, 21, 41
synthesis, vii, 1, 2, 3, 4, 7, 8, 9, 10, 11, 13, 14, 15, 17, 18, 20, 21, 22, 23, 24, 27, 28, 29, 30, 31, 33, 34, 35, 41, 42, 43, 44, 45, 46, 49, 51, 54, 55, 57, 59, 60, 61, 62, 63, 65, 66, 67, 68, 70, 71, 72, 73, 74, 75, 77

T

target, 10, 11, 14, 20, 24, 26, 31, 32, 41, 43, 50, 63, 64, 70, 72, 77
taxonomy, 71
techniques, 20
temperature, 39, 40, 45
thanatos, 61
threonine, 41
tissue, 12, 26, 27, 43, 64, 74
tobacco, 13, 16, 29, 32, 48, 50, 59, 69, 70, 72, 73
toxicity, 6, 18, 28, 53, 75
tracks, 30
trafficking, 14, 61, 65

transcription, 35
translocation, 26, 76
transport, 31, 33, 43
treatment, 7, 10, 13, 15, 18, 19, 22, 24, 27, 30, 32, 33, 36, 45, 61
triacylglycerides, 43
turgor, 71
turnover, 72

V

velocity, 30

W

water, 59
weight gain, 38
wild type, 9, 10, 39, 43, 76

X

xylem, 6, 39, 40, 75

Z

zinc, 22, 67